Quantum Mechanics in Everyday Life

Dr. Wilton L. Virgo, PhD

Rh

Rhodium Inc., Cambridge, MA

Quantum Mechanics in Everyday Life

BY
Dr. Wilton L. Virgo, PhD

Illustrated by Dr. Wilton L. Virgo, PhD

Rh

Second Edition 2014
Printed in the United States of America
ISBN 978-0-9909324-0-6
Library of Congress Control Number: applied for

Published by Rhodium, Inc.
P.O. Box 397073
Cambridge, MA 02139-7073 U.S.
www.quantumeveryday.com

Preface, 2nd Edition

Quantum mechanics is one of the most important branches of modern science. You may already know that quantum theory is central to physics because it describes the behavior of elementary particles and atoms. Did you also know that the same quantum theory is the foundation for all of chemistry? Quantum mechanics provides the mathematical framework for understanding the behavior of molecules, and gives deep insight into how and why chemical reactions between molecules take place. Every student of science learns at least something about quantum mechanics in the description of atomic and molecular behavior in the classroom or research laboratory. Professors and research scientists all over the world are currently performing sophisticated experiments and developing advanced theories in order to understand the complex quantum mechanical nature of atoms and molecules. What is the mathematical language of quantum mechanics? Is quantum mechanics at all relevant to everyday life?

This book addresses the question of how the seemingly abstract, mathematical concepts of quantum mechanics are the foundation of important technology in our modern society. The book explains how quantum mechanics is applied to modern, cutting-edge science and technology that directly affects our everyday lives. I wrote this book in order to bridge fundamental science and state-of-the-art technology that are both built upon quantum mechanical ideas.

The book is geared toward all who are curious about the dynamic, microscopic world, where probabilistic quantized particles tunnel through barriers while sending us clues about how the universe operates. Both undergraduate and graduate students in the sciences will benefit through engagement with ideas that go beyond the traditional classroom text to discover ways in which quantum unexpectedly appears in familiar places.

There are a large number of textbooks on quantum chemistry and physics. However, very few books provide students with the fundamental mathematical tools necessary to gain entry into both the cutting-edge research literature and the elegant, classic texts of chemical physics, while making connections to modern science and technology that have enhanced life in our society.

A thorough examination of the references cited in the text is necessary for mastering the fundamental concepts and mathematical formulas in the book. Reading these references will give you access to the relevant scientific literature and help you understand the basic ideas that are essential in learning quantum mechanics.

The first three chapters of the book will provide you with an understanding of the necessary mathematical concepts that underpin quantum mechanics. Your main goal in reading these chapters is to understand that the mathematical description of the microscopic world is profoundly different than the mathematical description of the macroscopic world that is familiar to most people. With these new mathematical tools in hand, you will learn how to apply the concepts of quantum mechanics to modern science and technology. The concepts illustrated by the examples in chapters 4-10 will upgrade and update your knowledge to a basic comprehension of how quantum mechanics is useful in everyday life.

The 2nd edition has been significantly updated and expanded. New material on the origins of quantum theory appears in chapter 1, with a discussion of Planck's blackbody equation and Einstein's quantum contribution. Chapter 2 now includes an introduction to the mathematics of statistical mechanics, while chapter 7 has been modified to include additional applications of quantum-based light sources. The complete text was typeset with LaTeX. All figures were produced using either Apple Pages [36], SAGE [80] or Google SketchUp [37].

In order to be competitive and engaged in our society of the 21st century and beyond, you must have at least a rudimentary knowledge of quantum mechanics for an understanding of how the modern world works. My goal in writing this book is to instill a desire for lifelong learning in the physical sciences, with quantum mechanics as the foundation of modern scientific thought.

Wilton L. Virgo, PhD
October 2014

Contents

Chapter 1

What is Quantum Mechanics?

In the early part of the 20th century, the scientific ideas associated with the emergence of quantum mechanics may have seemed to be esoteric. You may not have realized that the ideas behind quantum mechanics have shaped our global society and how we think about our world over the past 100 years. Quantum has completely changed our view of the world and the scientific laws that govern reality [13,98]. In fact, quantum theory has shaped our ideas of the postmodern world. You can find evidence of quantum mechanical ideas not just in science and technology, but also in art, literature, philosophy and business.

If you don't already know, the first question you should now be asking is, "what is quantum mechanics?" Simply put, quantum mechanics [27] is the mathematical description of the microscopic world of atoms, molecules and light. You should already be familiar with Newton's three laws of motion [59]. Newton's laws are applicable over a very wide range of size scale, from baseballs to planets and beyond. However, Newton's laws fail to describe objects at the size scale of atoms and molecules. The laws of quantum mechanics are different than those of Newtonian mechanics, and they provide us with a way of describing microscopic systems.

1.1   Discreteness

1.1.1   Planck: Quantization of Light

One of the key underlying principles of quantum mechanics is the idea of the discrete nature of energy. Prior to the development of quantum mechanics, electromagnetism was thought to have a continuous range of energy under Maxwell's classical electromagnetic theory [29]. This idea was completely upended when Max Planck proposed his quantum hypothesis [52, 60] in 1900, marking the beginning of the development of quantum theory. Planck hypothesized that energy comes in discrete packets, or quanta. When it comes to electromagnetic energy, these quanta are called photons. Planck said that a quantum of energy of is related to frequency, ν, multiplied by a constant, h, now called Planck's constant.

$$E = h\nu \tag{1.1}$$

or equivalently,

$$E = \hbar\omega \tag{1.2}$$

where $\omega = 2\pi\nu$ and $\hbar = h/2\pi$. Planck's constant is very small, 6.626×10^{-34} J · s, but it is not zero. This magnitude of h is very important because the frequency of visible light is in the range of 10^{14} Hz, so the corresponding energies in Planck's formula are small (10^{-19} J). In comparison, the energy gained from consuming an average candy bar is one million Joules (10^6 J), so one would not expect quantum behavior in everyday life. $E = h\nu$ only becomes important at the atomic and molecular level, but we will see in later chapters that macroscopic technology based on the behavior of atoms and molecules would not be possible without Planck's contribution to science.

Planck was able to show that matter, which is made up of atoms and molecules, can only absorb or emit discrete quanta of electromagnetic radiation. Planck verified his hypothesis by mathematically predicting the relationship between the intensity and frequency of electromagnetic radiation emitted by a blackbody. A blackbody is any object that completely absorbs and re-emits all of the electromagnetic radiation that interacts with the body, without transmitting, reflecting or scattering that electromagnetic ra-

diation. Understanding blackbody behavior is very important, because our sun, or any other star in the universe for that matter,
can be modeled as a blackbody using Planck's theory. Planck's
quantum blackbody law is,

$$d\rho(\nu, T) = \rho_\nu(T)d\nu = \frac{8\pi\nu^2}{c^3}\frac{h\nu}{e^{\frac{h\nu}{k_BT}} - 1}d\nu \qquad (1.3)$$

where ρ is the radiant energy density between frequency ν and
$\nu + d\nu$, T is the temperature, $c = 299792458$ m/s is the speed
of light and $k_B = 1.38 \times 10^{-23}$ J/K is the Boltzmann constant.
Since the relationship between the speed of light, wavelength and
frequency is $c = \lambda\nu$ [88], we can also write the quantum blackbody
equation in terms of wavelength,

$$\rho_\lambda(T)d\lambda = \frac{8\pi hc}{\lambda^5}\frac{1}{e^{\frac{hc}{\lambda k_BT}} - 1}d\lambda \qquad (1.4)$$

Planck's quantum formula gives a perfect fit to experimental data

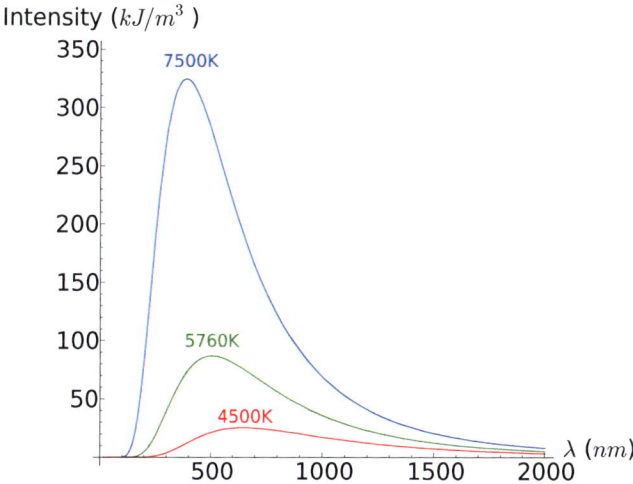

Figure 1.1: Energy density of Planck's blackbody radiation.

collected from hot objects, like a burner on an electric stove, or
the sun. Figure 1.1 shows a plot of Planck's quantum blackbody
radiation law as a function of wavelength for several temperatures.

Planck's law can be differentiated with respect to wavelength and set equal to zero to give the relationship between temperature and the maximum wavelength of blackbody emission:

$$\frac{d}{d\lambda}[\rho_\lambda(T)] = \lambda_{max}T = 2.90 \times 10^6 \text{ nm} \cdot \text{K} \tag{1.5}$$

This relationship where the product of the peak wavelength and temperature is equal to a constant is called Wein's displacement law. When intensity is plotted as a function of wavelength, it can be used to determine the temperature of any hot object. Our sun's wavelength peaks near 500 nm. Plugging this λ_{max} into equation (1.5) gives a surface temperature of 5800 K. A little knowledge of quantum mechanics allows you to determine the surface temperature of our sun, just by observing the sun's peak color!

Although no real object behaves as a perfect blackbody, the canonical blackbody is a uniformly heated, hollow cavity which emits electromagnetic radiation from a small hole. Planck's law can also be integrated by substitution to find the finite total electromagnetic radiation energy per unit volume inside any blackbody cavity:

$$\int \rho_\nu(T)d\nu = \frac{8\pi h}{c^3}\int_0^\infty \frac{\nu^3}{e^{\frac{h\nu}{k_BT}}-1}d\nu = \frac{8\pi h}{c^3}\left(\frac{k_BT}{h}\right)^4\int_0^\infty \frac{x^3}{e^x-1}dx \tag{1.6}$$

with $x = (h\nu)/(k_BT)$. The integral can be evaluated as,

$$\int_0^\infty \frac{x^3}{e^x-1}dx = \frac{\pi^4}{15} \tag{1.7}$$

The resulting Stefan-Boltzmann law shows that the total energy density for blackbody radiation is proportional to the fourth power of the temperature,

$$\rho(T) = \sigma T^4 \tag{1.8}$$

where $\sigma = \frac{8\pi^5 k_B^4}{15c^3h^3}$ is the Stefan-Boltzmann constant.

$$\sigma = \frac{8\pi^5 k_B^4}{15c^3h^3} = 7.565 \times 10^{-16} \frac{\text{J}}{\text{m}^3\text{K}^4} \tag{1.9}$$

However, you can multiply equation (1.8) by $\frac{c}{4}$, and that describes the rate of emission from a surface at temperature T. That gives the Stefan-Boltzmann constant for surface blackbody emission,

$$\sigma = 5.67 \times 10^{-8} \frac{\text{W}}{\text{m}^2\text{K}^4} \tag{1.10}$$

An understanding of blackbody radiation also provided the main supporting experimental evidence for the Big Bang theory of the creation of the universe. Twenty years ago, George Smoot and John Mather [50] used NASA's COBE (Cosmic Background Explorer) satellite to measure the cosmic microwave background (CMB) radiation, which is the leftover energy from the Big Bang. The CMB is the furthest back in time that we can observe our 14 billion year old universe using light. Imprinted on the CMB are the seeds of galaxies that we can observe today. Planck's theory of blackbody radiation gave a great fit to the experimental microwave background radiation data observed by COBE, corresponding to a temperature of 2.7 K.

The foundation of Planck's blackbody radiation theory is his quantum hypothesis, $E = h\nu$. As quantum theory developed, it was realized that atoms and molecules have well-defined, quantized energy levels. Since photons are discrete packets of energy, atoms and molecules can absorb or emit energy in discrete units of these quantized bundles of energy. The discrete photon energy drives transitions between the atomic and molecular energy levels. This idea of the discrete nature of matter and energy is one of the most important and recognizable concepts in quantum mechanics. Planck's formula, $E = h\nu$, is the elegant mathematical description of the concept of quantum discreteness. One of the reasons why Planck's quantum hypothesis was such a radical change in thinking is that it could not be derived from any scientific theory that existed in western society in 1900, such as Newton's laws of motion or Maxwell's classical theory of electromagnetism. The hypothesis just has to be accepted as truth based on observations of nature, like blackbody radiation from the sun or the microwave energy left over from the Big Bang.

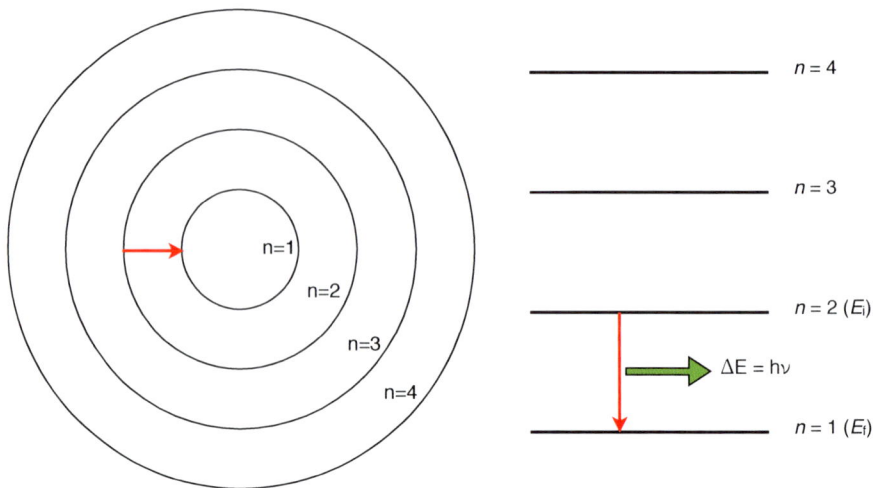

Figure 1.2: The Bohr model of the hydrogen atom.

1.1.2   Bohr: Quantization of Matter

In 1913, Niels Bohr created a model for the quantum structure of atoms. Bohr's model [5–7, 40] was revolutionary because it took Planck's quantum hypothesis and applied the discreteness concept to individual atoms. In Bohr's model of the atom, the electrons in the atom can only occupy certain discrete energy levels. In other words, the energy of the atom can only take on certain allowed values. The electronic energy of the atom can change when one of the electrons makes a transition from one discrete energy level to the next, but this must be accompanied by the simultaneous absorption or emission of a photon. A photon is absorbed upon a transition to a higher energy state, or emitted during a transition to a lower energy state. As an electron jumps from an initial state, E_i, to a final state, E_f, a photon is either absorbed or emitted with a quantum of energy that is equal to the energy difference between the two states connected by the transition.

$$E_i - E_f = h\nu \qquad (1.11)$$

or equivalently,

$$\Delta E = h\nu \tag{1.12}$$

In Bohr's model, the possible values for the energy levels of a one-electron atom are,

$$E_n = -\frac{\mu(Ze^2)^2}{2\hbar^2 n^2} \tag{1.13}$$

where n is the set of positive integers ($n = 1, 2, 3, 4, 5, ...$), Z is the atomic number, e is the fundamental unit of charge and

$$\mu = \frac{m_n m_e}{m_n + m_e} \tag{1.14}$$

where m_n is the nuclear mass and m_e is the mass of the electron. Bohr's elegant model works for the hydrogen atom, all other one-electron atoms, and highly excited states of atoms called Rydberg states [43].

As one of the important keys to his model, Bohr made a postulate that the angular momentum of the electron in the atom can only increase or decrease by discrete amounts. The electron can only jump from one quantum state to the next if the angular momentum of the electron changes by units of $\hbar$. In other words, the orbital angular momentum of the electron is quantized. We will return to a discussion of quantized angular momentum [9, 22, 65, 97] in chapter 3.

1.1.3 Einstein's Quantum Approach

In 1917, Albert Einstein built upon the quantum theories of Planck and Bohr, using statistical thermodynamics to derive Planck's quantum blackbody radiation law. Let's come back to the idea of a blackbody cavity. The walls of the cavity are composed of atoms, and Bohr's theory tells us that these atoms can exist in quantized states labeled by energies, E_n. If more than one quantum state has the same energy, this results in degenerate energy levels. We denote the number of degenerate levels by the label g. The description of a macroscopic system in terms of an assembly of quantized microscopic entities requires a statistical mechanics approach. Using the Boltzmann distribution, we can determine the ratio of the number

7

of atoms in one state (with label m) to the number of atoms in another state (labeled n):

$$\frac{N_m}{N_n} = \frac{g_m e^{\frac{-E_m}{k_B T}}}{g_n e^{\frac{-E_n}{k_B T}}} = \frac{g_m}{g_n} e^{-(E_m - E_n)/k_B T} \qquad (1.15)$$

Figure 1.3 is a diagram of a two-level system with energies E_1 and E_0, such that $E_1 > E_0$. Bohr told us that photon absorption and emission drives the transitions between any two levels. We want to mathematically describe the rate of transitions between the two levels. A rate is just the number of transitions per unit time. It depends on the number of atoms in the initial state, and the intensity of the radiation in the cavity, $\rho(\nu, T)$. The transition rate for photon stimulated absorption from E_0 to E_1 can be written as,

$$R_{0 \to 1} = N_0 B_{01} \rho(\nu, T) \qquad (1.16)$$

where B_{01} is the absorption coefficient. For the opposite process of emission, there are two mechanisms for transition from the upper level to the lower one. Stimulated emission can induce transitions to the lower level, with coefficient B_{10}. Spontaneous emission occurs at a rate independent of any photons that may be present, with a coefficient A_{10}:

$$R_{1 \to 0} = N_1 [A_{10} + \rho(\nu, T) B_{10}] \qquad (1.17)$$

At thermal equilibrium, the rate of absorption must equal the rate

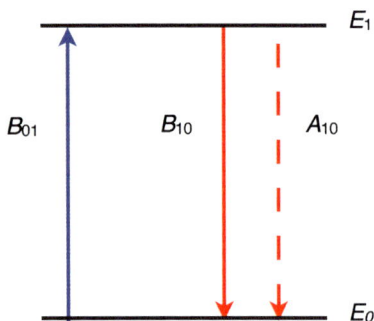

Figure 1.3: Transitions in a two-level atom.

of emission,

$$R_{0 \to 1} = R_{1 \to 0} \tag{1.18}$$

so that,

$$\frac{N_0}{N_1} = \frac{A_{10} + \rho(\nu, T)B_{10}}{B_{01}\rho(\nu, T)} = \frac{g_0}{g_1} e^{-(E_0 - E_1)/k_B T} \tag{1.19}$$

Solving equation (1.19) for $\rho(\nu, T)$, we have,

$$\rho(\nu, T) = \frac{A_{10}/B_{10}}{\frac{g_0 B_{01}}{g_1 B_{10}} e^{(E_1 - E_0)/k_B T} - 1} \tag{1.20}$$

Since the rate of stimulated absorption should equal the rate of stimulated emission,

$$\frac{g_0 B_{01}}{g_1 B_{10}} = 1 \tag{1.21}$$

The ratio of the rate of spontaneous emission to the rate of stimulated emission is,

$$\frac{A_{10}}{B_{10}} = \frac{8\pi\nu^2}{c^3}(E_1 - E_0) \tag{1.22}$$

Finally, we bring back Planck's quantum hypothesis, $\Delta E = h\nu = E_1 - E_0$. Pulling this all together gives Planck's quantum blackbody law!

$$\rho(\nu, T)d\nu = \frac{8\pi\nu^2}{c^3} \frac{h\nu}{e^{\frac{h\nu}{k_B T}} - 1} d\nu \tag{1.23}$$

1.2   Probability

Another central concept in quantum mechanics is probability. Every observable, or measurable quantity, in quantum mechanics has an associated probability. This is key, because one can't always say that a particular observable in quantum mechanics is definite. The best one can do is compute the probability that a certain behavior will be observed. Before quantum mechanics revolutionized science, the idea of determinism was one of the fundamental assumptions in classical physics. It means that if the motion of any object in the universe is known at a given time, its motion before or after that time could be completely determined using Newton's laws. In the quantum world where probability governs outcome, this classical

9

idea of determinism completely falls apart [19]. Let's take a look at an example with polarizing glasses and polarized light to show that determinism does not apply in the quantum world.

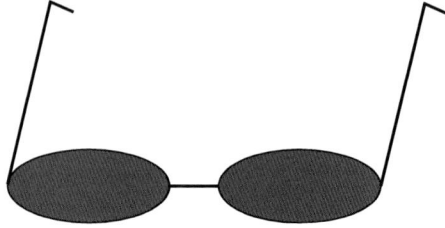

Figure 1.4: Polarizers that might be used in 3D glasses for viewing a 3D movie.

If you recently saw a 3D movie, you have used polarizers. These are the lenses in the special glasses that you use to watch the 3D movie (figure 1.4). The polarizers give the illusion of depth (the third dimension), because each of your eyes sees a slightly different image. Your brain interprets these two images as a 3-dimensional picture. The 3D movie glasses have two polarizing lenses which have their direction of polarization oriented 90 degrees with respect to one another. The next time you go to see a 3D movie, take two glasses and put a lens from one pair of glasses directly in front of the other. Next, look through the two lenses at a light source, like a light bulb. Now rotate one lens with respect to the other, and you will see that the amount of light passing through the two lenses will increase and decrease as you rotate the lenses. In rotating the lenses, you are rotating the polarization direction of the glasses. If you have polarized light going through a polarizer,

and if the polarizer is physically oriented parallel to the polarization direction of the light, then all the light will pass through the
polarizer. If the polarizer is rotated so that it is perpendicular to
the polarization of the light, then no light will get through. If the
polarizer is rotated at 45 degrees, then the intensity of light will be
cut in half. The fraction of light transmitted through the polarizer
can be generally described as $\cos^2 \theta$, where θ is the angle between
the orientation of the polarizer and the electric field of the light
(figure 1.5). In one type of 3D movie theater, two projectors are
used that have polarizing lenses in front of them that match the polarization directions of the glasses that you wear. The polarization
is set such that each of your eyes only sees the image transmitted
by one of the projectors that matches the polarization of the lens
in front of your eye.

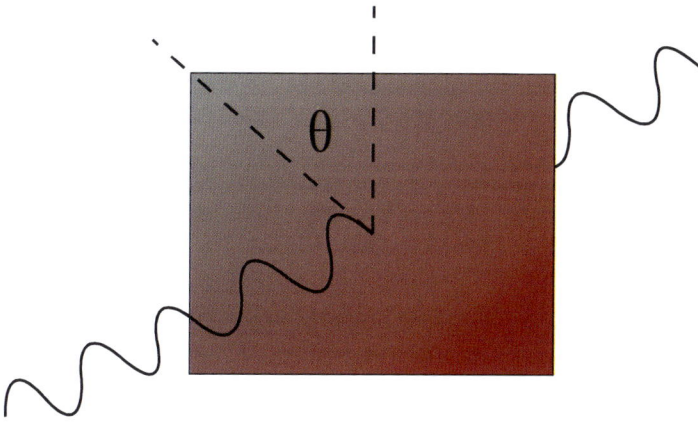

Figure 1.5: Electric field of light passing through a polarizer.

Now let's do a thought experiment and use Planck's quantum
hypothesis to see how a single photon would pass through a polarizer. A 100 W light bulb puts out about 10^{20} photons/sec. Let's
imagine that we could turn the intensity of a light bulb way down
so that we could detect one photon at a time as it passes through
the polarizer (figure 1.5). Only one of two things can happen to
a single photon. Either the photon is transmitted through the po-

larizer, or it isn't. However, since the photon lives in a quantum world, the *probability* of the photon passing through the polarizer varies as $\cos^2 \theta$. The quantum world is probabilistic.

1.2.1 Wave-Particle Duality

Next, let's examine the concept of wave-particle duality. In classical physics, energy can exist in two forms. It can either be a moving particle, or a wave. It must be one of the two. In the study of classical physics, one knows that a wave can't be localized at a particular point in space, but a particle can be localized at a particular coordinate in 3D space. Planck's quantum hypothesis suggested that photons, as packets of energy, might display characteristics normally associated with particles. Einstein's explanation of the photoelectric effect [23, 88] in 1905 showed that even though light can exhibit wavelike behavior (diffraction, reflection, refraction, interference), photons of light can also display particle-like behavior in the ejection of electrons from metal surfaces.

1.2.2 deBroglie Wavelength

In 1924, Louis deBroglie [20] proposed a new, groundbreaking idea that blurred the line between the concept of a wave and a particle imposed by classical physics. deBroglie hypothesized that if light can exhibit both wave- and particle-like properties, then perhaps microscopic matter might also exhibit wave- and particle-like characteristics. deBroglie came up with an equation that related the momentum of a particle to its wavelength.

$$p = \frac{h}{\lambda} \tag{1.24}$$

Note the appearance of Planck's constant, which has appeared as a part of every group of equations we have examined so far in this chapter.

Let's use deBroglie's formula to calculate the wavelength of a baseball, and compare it to the deBroglie wavelength of an electron. The mass of a baseball is about .15 kg, and a 90 mph fastball is moving at a speed of about 40 m/s.

$$\lambda = \frac{h}{p} = \frac{h}{mv} = \frac{6.626 \times 10^{-34}\,\text{J}\cdot\text{s}}{(.15\,\text{kg})(40\,\text{m/s})} \tag{1.25}$$

$$\lambda = 1 \times 10^{-34} \, \text{m}$$

since $1\text{J} = 1\frac{\text{kg·m}^2}{\text{s}^2}$. The deBroglie wavelength of a baseball is negligibly short. Wavelike properties are not observable for baseballs. This is a good result! Calling a strike or a ball at a major league game is difficult enough without having to take into account the wavelength of baseballs.

For an electron, the mass is 9.11×10^{-31} kg, and the velocity of an electron accelerated in an electric field may be on the order of 10^6 m/s.

$$\lambda = \frac{h}{p} = \frac{h}{mv} = \frac{6.626 \times 10^{-34} \, \text{J} \cdot \text{s}}{(9.11 \times 10^{-31} \, \text{kg})(10^6 \, \text{m/s})} \quad (1.26)$$

$$\lambda = 7 \times 10^{-10} \, \text{m}$$

This is an important result, because the distance between atoms in a crystal is on the order of 10^{-10} m, close to our calculated deBroglie wavelength for the electron. Diffraction can occur when the spacing between scatterers is on the order of the wavelength. This is the basis for the use of electron diffraction in determining the structure of crystals. Electron diffraction studies take advantage of the wavelike properties of electrons in the same way that one would use x-rays (high energy photons) to determine crystal structure using x-ray diffraction (in fact, x-rays are usually generated by accelerating electrons in the laboratory). The calculation of the deBroglie wavelength of an electron also shows that the wavelength is on the order of the size of a typical bond distance ($\approx 1\text{Å} = 10^{-10}$ m) between atoms in a molecule. One would say that the electron is *delocalized* over the molecule as a wave.

Think about what would happen if you could take a gas-phase molecule at room temperature, and slow it down to low velocity by cooling the molecule. According to deBroglie's formula, as the molecule is cooled and slowed, the deBroglie wavelength becomes larger and larger! If the molecule can be cooled and slowed down enough, the deBroglie wavelength will increase to a length that is significantly larger than the typical distance between two colliding, reacting molecules. If this can be done experimentally, it opens up a new mechanism for chemical reactions. Instead of having a hard collision between reacting molecules, they may interact and react at relatively large distances via quantum tunneling (which will be discussed toward the end of this chapter) due to their long deBroglie wavelengths. Normally, one would heat a mixture of chemicals to increase the rate of a reaction. One might not expect any chemical reactions to take place at the low temperatures needed to make the long deBroglie wavelengths discussed here, because the molecular collision energies are too low to overcome any activation barriers to reaction. However, the long deBroglie wavelengths of low-velocity, ultracold molecules open up new quantum channels for chemistry. There is a large effort by many research groups around the world to study cold atoms and molecules in order to observe and test these kinds of quantum effects at low temperature [30, 31, 85].

deBroglie's formula is directly connected to the quantum concept of probability. The wavelike nature of microscopic matter means that quantum particles cannot be localized in space. One is forced to think about these small particles in terms of a probability amplitude instead of a well-defined object in 3D coordinate space.

1.3   Uncertainty and Matrix Mechanics

A third key concept in quantum mechanics is the idea of uncertainty. Under Newton's laws of motion, the motion of any object could be determined to any desired precision if one knew the properties (mass, velocity) of that object at a given time. Quantum mechanics says something very different. According to the theory, one can never be completely certain about the motion of a quantum object. Werner Heisenberg proposed a formula that said one can't simultaneously measure the position and momentum of a microscopic particle to any arbitrary precision. There is a limit to the precision of the measurement. This limit is Planck's constant. If you want to measure the position of the particle more precisely, you necessarily have to sacrifice the certainty with regard to the momentum, and vice versa. Heisenberg's Uncertainty Principle [40, 52, 59] can be expressed as,

$$xp_x - p_x x = i\hbar \tag{1.27}$$

where x is the position of a quantum particle, and p_x is its momentum. This form of the Uncertainty Principle shows how Heisenberg took the first steps toward developing a new mathematical quantum language. His work was innovative because rather than thinking in terms of normal observable quantities like the classical orbit of electrons around a nucleus, he used the ideas of discreteness, probability and uncertainty to form a more abstract language grounded in the algebraic theory of matrices. In the matrix formulation of quantum mechanics, both x and p_x can be represented by matrices. The important thing to recognize in equation (1.27) is that the quantity, $xp_x - p_x x$, is not zero. The commutative property of multiplication of the set of real numbers does not apply in this case. However, we know that matrix multiplication is not necessarily commutative. In the language of matrix mechanics, the position and momentum of a microscopic particle do not commute. We can rewrite Heisenberg's Uncertainty Principle in terms of the commutation relationship between p_x and x, where,

$$[x, p_x] = xp_x - p_x x = i\hbar \tag{1.28}$$

or as the following relation [2, 40, 61],

$$\Delta x \Delta p_x \geq \hbar/2 \tag{1.29}$$

In other words, the product of the uncertainty in position and uncertainty in momentum has a limit, $\hbar/2$.

1.3.1 Heisenberg, Born, Pauli and Dirac

Werner Heisenberg, Max Born, Wolfgang Pauli and Paul Dirac made great advances [51] with the use of matrix mathematics in the development of quantum mechanics. Max Born recognized that Heisenberg's mathematical formulation really was the language of matrices. Pauli showed that Heisenberg's matrix mechanics could be used to explain and predict atomic spectra. Dirac combined the mathematics of matrix mechanics and electromagnetic theory into quantum electrodynamics. These advances showed that any problem in quantum mechanics can be solved using the algebra of matrices. In chapter 3, we will discuss how matrices can be used to represent angular momentum operators, the tools one can use to calculate the energy level structure of atoms and molecules.

1.3.2 Scanning Tunneling Microscopy (STM)

One important example of modern technology which is based on the quantum concepts of probability and uncertainty is Scanning Tunneling Microscopy, or STM. Using the STM technique, one can obtain an image of a surface with atomic resolution. STM works because of quantum tunneling. The tunneling microscope (figure 1.6) consists of a conductor with a fine tip that is brought very close to a metal or semiconductor surface, all inside a very low-pressure vacuum chamber. A voltage is applied to the tip, while the surface is grounded. Electrons at the tip would like to flow onto the surface, but the vacuum of empty space separates the tip from the surface. However, due to the Uncertainty Principle and the wave-nature of electrons, one can't be absolutely sure exactly where the electrons are, or where they will be. The electrons can actually tunnel across the vacuum gap from the conducting tip to the surface. If one measures the current produced by the tunneling electrons as the probe tip is moved over the surface, one can construct an image completely based on the phenomenon of quantum tunneling.

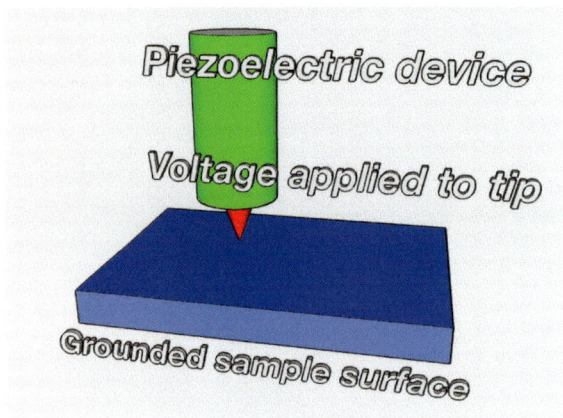

Figure 1.6: Design for a Scanning Tunneling Microscope (STM). Electrons tunnel from the tip to the surface.

This is a real-world scientific application of probability, uncertainty and the wavelike nature of electrons. From Heisenberg's Uncertainty Principle, you now know that an electron does not have a well-defined trajectory in space. In measuring the tunneling current, you also know the electron started at the tip and you know it ended up at the surface, but you have no idea how it got from one place to the other. Due to the wavelike nature of the electron described by deBroglie's formula, the best you can do is to determine the probability of the electron tunneling from tip to surface.

1.4   Summary

Discreteness, probability and uncertainty are the hallmarks of quantum mechanics. These foundations of quantum are radically different than the fundamental ideas that the average person may have about the macroscopic world. Quantum theory may appear limiting at first glance because only certain discrete energies are allowed to exist and one can't make measurements to any arbitrary precision. One can only predict what will happen with a certain probability. This is very different from our experience in the macroscopic world because we have to think about physical concepts such as size, position, momentum, velocity and acceleration in a totally different way with quantum mechanics. You can't use your intuition about the macroscopic world and apply it to atoms and molecules. However, don't be fooled into thinking that quantum mechanics has no direct effect on everyday life and activities. Even though quantum mechanics is characterized by probability and uncertainty, technology based on quantum mechanics is being used to observe objects at the far-reaches of the universe through radio astronomy techniques. Atomic clocks and the global positioning system (GPS) would not be possible without quantum mechanics, and these technological tools provide us with the most precise measurements of time and position known to humankind. It is difficult to find a part of the modern world that has not benefited from knowledge of quantum mechanics. Quantum theory is pushing the limits of computer technology and advanced sensors, bringing new insights into alternative green energy, enhancing our fundamental understanding of global warming and climate change, and is responsible for advances in biomedical science.

In 2014, the widely available technology based on quantum mechanical principles is likely used by you on a daily basis, although you may not realize it. Since much of modern technology is based on the fundamental nature of atoms and molecules, an understanding of quantum mechanics is becoming increasingly important for anyone who wishes to understand how devices based on modern technology work. We will see examples of the latest quantum-based technology in chapters 4-10. Prior to going through these examples of quantum in technology, we must introduce the mathematical formalism of quantum mechanics in order to make the connection between the mathematical foundation and everyday applications of quantum in the 21st century.

Chapter 2

The Mathematics of Quantum Mechanics

You already know that mathematics is the language of science. Every scientific discipline is based on a well-defined set of mathematical laws. Quantum mechanics is simply the fundamental language of chemistry and physics. Almost any phenomenon in chemistry or physics that has to do with molecules or atoms can be described using the mathematical language of quantum mechanics. One of the great things about science is that the same mathematical language of quantum mechanics is used by all scientists and students in every country. It does not matter whether you are American, Ghanaian or French. When it comes to quantum mechanics, the mathematical language is the same. You should feel empowered by the fact that you are able to communicate quantum mechanical ideas with anyone in the world, and comprehend the basic ideas.

In this book, the mathematical language that we will use to describe quantum mechanics is the algebra of matrices. Instead of using differential equations that require the use of calculus, we will take advantage of the fact that matrices represent physical quantities in quantum mechanics. The position, velocity, linear momentum, angular momentum and energy of an atom or molecule can be represented by a matrix. You only need to know the basic algebra of matrices, complex numbers and probability to master the concepts in this book.

2.1 Key Postulates in Quantum Mechanics

While Newton's laws (classical mechanics) apply to everything from planets to baseballs, they fail when dealing with atoms and molecules. Quantum mechanics is the language of the microscopic world. Quantum has explained everything from chemical bonding and reactions and high-energy physics to lasers, MRI and microelectronics that we use every day.

2.1.1 The Wavefunction and Probability

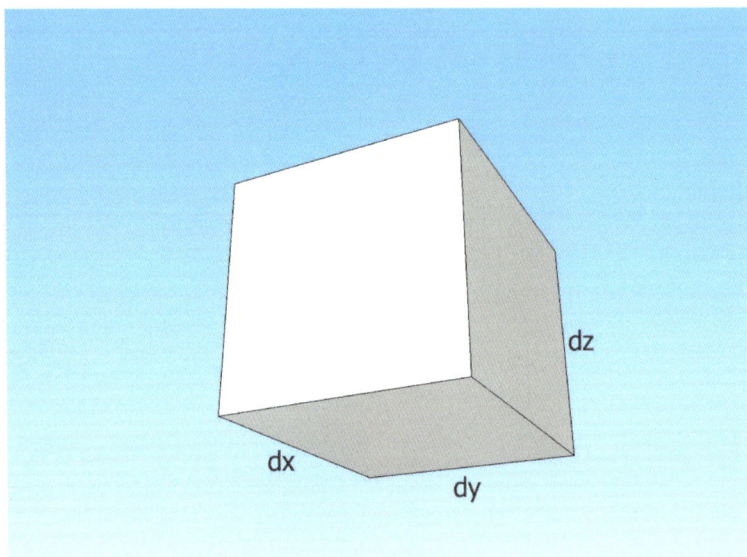

Figure 2.1: A 3D quantum box.

The state of any quantum mechanical system (atom or molecule) can be described by a function called the wavefunction, $\Psi(r, t)$, which is a function of coordinates of particles and time. The expression

$$\Psi(r, t)^* \Psi(r, t) dx dy dz \tag{2.1}$$

tells you the probability that a quantum particle lies somewhere in volume $dx dy dz$ at some coordinate r at time t. This is also called the Born interpretation [8] of quantum mechanics, where

$\Psi^*\Psi = |\Psi^2|$ is recognized as a probability density, $P(r,t)$, for a particle located at a point, r, at time, t.

The total probability is equal to one,

$$\int P(r,t)d\tau = \int |\Psi(r,t)|^2 d\tau = 1 \tag{2.2}$$

since the particle must be located somewhere in the box with volume $d\tau = dxdydz$. This is called *normalization*. All wavefunctions in quantum mechanics are *normalized.*

2.1.2 The Eigenvalue Equation

In quantum mechanics, an *operator* represents a set of mathematical instructions for carrying out some action or function. An *observable* is a variable that can be measured (i.e. energy, momentum). If an operator operates on a function to give the same function back multiplied by a constant, this is called an eigenvalue equation. All eigenvalue equations look like this:

$$\hat{O}\Psi = a\Psi \tag{2.3}$$

where Ψ is an eigenfunction of the $\hat{O}$ operator, and a is a constant called an eigenvalue. When you measure an observable associated with an operator, the only values that will be observed are eigenvalues of that operator. For example, when you measure the energies of an atom or molecule with the energy operator (called the Hamiltonian), the observed energies, E, must be eigenvalues of the Hamiltonian, $\hat{H}$.

$$\hat{H}\Psi = E\Psi \tag{2.4}$$

2.1.3 Mean Values for Observables

Let's say we have a system that can be described by a normalized wave function. The average value of an observable associated with that system is,

$$\langle a \rangle = \int_{-\infty}^{\infty} \Psi^* \hat{O} \Psi d\tau \tag{2.5}$$

21

For example, if we want to know the average, or expectation value for the position of a quantum particle,

$$\langle r \rangle = \int r P(r,t)d\tau = \int \Psi^*(r,t)r\Psi(r,t)d\tau \qquad (2.6)$$

These equations tell us that you have to think about position, energy, velocity, acceleration, momentum and size totally differently. These concepts that are familiar in Newtonian mechanics are given a weird, abstract mathematical description in quantum mechanics.

2.2 Properties of the Wavefunction

2.2.1 Wave Superposition

Just like waves on the ocean crashing into one another and interfering constructively and destructively, quantum waves can be superimposed mathematically.

$$\psi = c_1\psi_1 + c_2\psi_2 + c_3\psi_3 \qquad (2.7)$$

where the lowercase ψ is used to represent a time-independent wavefunction. You can superimpose as many waves as you like,

$$\psi = \sum_n c_n\psi_n \qquad (2.8)$$

where n can run from 1 all the way up to ∞. The c_n are coefficients that could be complex numbers. The ψ_n could be wavefunctions for an atom or molecule.

2.2.2 Overlap Integrals

Imagine that you have an electron in a box divided into two sections, F and G, by a wall that the particle cannot pass through. You can think of this as dividing the wavefunction into two waves that can't overlap, $c_F\psi_F$ and $c_G\psi_G$, where,

$$\psi = c_F\psi_F + c_G\psi_G \qquad (2.9)$$

Since ψ is normalized,

$$\int_{box} |\psi|^2 d\tau = |c_F|^2 + |c_G|^2 + c_F^*c_G \int_{box} \psi_F^*\psi_G d\tau + c_G^*c_F \int_{box} \psi_G^*\psi_F d\tau = 1 \qquad (2.10)$$

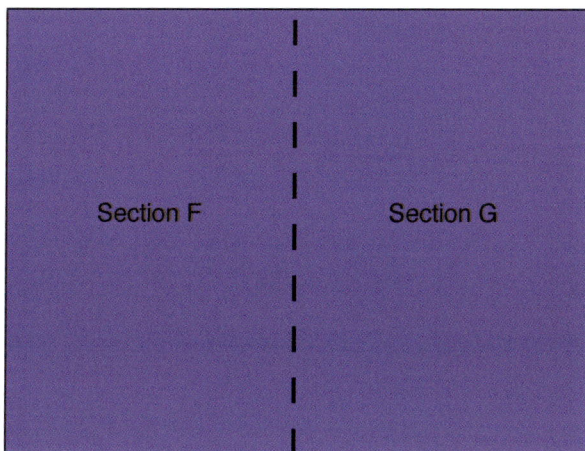

Figure 2.2: A quantum box divided into two sections, F and G.

The probability of finding the electron in section F is,

$$\int_F |\psi|^2 d\tau = |c_F|^2 \qquad (2.11)$$

while the probability of finding the electron in section G is,

$$\int_G |\psi|^2 d\tau = |c_G|^2 \qquad (2.12)$$

If you were to open the box to observe the electron, the wave-function would collapse, and you would find the electron in either F or G. The electron cannot be in both F and G simultaneously. The electron cannot be in neither F nor G simultaneously. In mathematical language, the following must be true:

$$|c_F|^2 + |c_G|^2 = 1 \qquad (2.13)$$

$$\int_{box} \psi_F^* \psi_G d\tau = \int_{box} \psi_G^* \psi_F d\tau = 0 \qquad (2.14)$$

23

The integrals in equation (2.14) are called *overlap* integrals. In this case, the overlap integrals vanish because ψ_F and ψ_G don't overlap, they are *orthogonal* because of the rigid wall in the middle of the box. Before the box is opened, you can only think in terms of the *waving* probability of finding the particle in one of the sections. If you open the box and observe the electron in section F, the probability of finding the electron in section G instantaneously drops to zero. If the electron is observed in section G, the probability of finding the electron in section F instantaneously drops to zero. The wavefunction has collapsed. Overlap integrals like these could describe the overlap of atomic orbitals on two different atoms coming together to form a diatomic molecule.

2.2.3 Orthonormality

We now understand overlap integrals, normalization, and orthogonality. Let's develop our quantum language further. Divide our box into many sections, where the total wavefunction is a superposition of individual wavefuctions,

$$\psi = \sum_n c_n \psi_n \qquad (2.15)$$

Orthogonality tells us that,

$$\int \psi_m^* \psi_n d\tau = 0, (m \neq n) \qquad (2.16)$$

Normalization tells us that,

$$\int \psi_n^* \psi_n d\tau = 1 \qquad (2.17)$$

We can combine orthogonality and normalization (this is called *orthonormality*) to give,

$$\int \psi_m^* \psi_n d\tau = \delta_{mn} \qquad (2.18)$$

where δ_{mn} is called the Kronecker delta. $\delta_{mn} = 0$ if m $\neq$ n and $\delta_{mn} = 1$ if $m = n$.

2.3 Dirac Notation: A New Language

We can simplify the quantum integral language using Dirac notation. The following expression will help you translate our ideas of wave superposition, overlap integrals and orthonormality from integral language to Dirac notation.

$$\left\langle m \left| \hat{H} \right| n \right\rangle = \int \psi_m^* \hat{H} \psi_n d\tau \qquad (2.19)$$

where the operator $\hat{H}$ is sandwiched in between a 'bra' $\langle|$ and 'ket' $|\rangle$. $\langle m|$ is a *bra* representing the complex conjugate of wavefunction ψ_m^*. $|n\rangle$ is a *ket* representing the wavefunction ψ_n.

Let's say that the operator in between the bra and ket in equation (2.19) is changed to "multiply by one". In this case,

$$\langle m \,|\, n \rangle = \int \psi_m^* \psi_n d\tau \qquad (2.20)$$

We can take the complex conjugate of this equation to give,

$$\langle m \,|\, n \rangle^* = \int (\psi_m^*)^* (\psi_n)^* d\tau = \int \psi_m \psi_n^* d\tau \qquad (2.21)$$

$$= \int \psi_n^* \psi_m d\tau = \langle n \,|\, m \rangle$$

We can translate normalization from integral language to Dirac notation as follows:

$$\int \psi_n^* \psi_n d\tau = 1 \qquad (2.22)$$

$$\langle n \,|\, n \rangle = 1$$

Translating orthogonality from integral language to Dirac language gives,

$$\int \psi_m^* \psi_n d\tau = 0, (m \neq n) \qquad (2.23)$$

$$\langle m \,|\, n \rangle = 0$$

Translating orthonormality from integral language to Dirac language gives,

$$\int \psi_m^* \psi_n d\tau = \delta_{mn} \qquad (2.24)$$

$$\langle m \,|\, n \rangle = \delta_{mn}$$

2.4   The Vector Representation

A vector has a magnitude and a direction in space. Physical quantities such as velocity and momentum can be represented by vectors. Two vectors ($\mathbf{F}$ and $\mathbf{G}$) can be added together in the following way to give a new vector, $\mathbf{H}$:

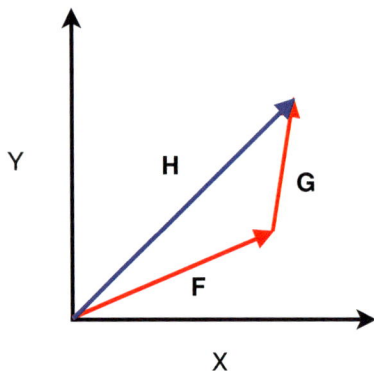

Figure 2.3: Addition of two vectors.

From basic geometry, a vector, $\mathbf{v}$, in 3D space can be expanded using a set of unit-length orthonormal basis vectors, $\mathbf{a}_x, \mathbf{a}_y, \mathbf{a}_z$ that point along the x, y and z-axes of a Cartesian coordinate system and a set of coefficients, c_x, c_y, c_z, which are the *components*, or *projections* of vector **v** along the Cartesian axes.

$$\mathbf{v} = c_x \mathbf{a}_x + c_y \mathbf{a}_y + c_z \mathbf{a}_z \tag{2.25}$$

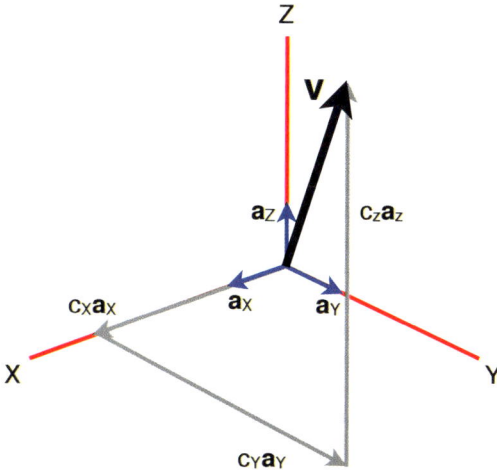

Figure 2.4: A 3D vector.

In quantum mechanics, we can generalize this vector idea to a space of infinite dimension.

$$\mathbf{v} = \sum_n c_n \mathbf{a}_n \tag{2.26}$$

In the same way we can write,

$$|\psi\rangle = \sum_n c_n |n\rangle$$

We can then write,

$$\mathbf{a}_m \cdot \mathbf{a}_n = \delta_{mn} \tag{2.27}$$

which is analogous to,

$$\langle m \,|\, n \rangle = \delta_{mn} \tag{2.28}$$

You can use this vector analogy to *project* out one of the expansion coefficients by using a dot product with one of the basis vectors.

$$\mathbf{v} \cdot \mathbf{a}_m = \sum_n c_n \mathbf{a}_n \cdot \mathbf{a}_m = \sum_n c_n \delta_{nm} = c_m \qquad (2.29)$$

Analogously,

$$\langle m \,|\, \psi \rangle = \sum_n c_n \langle m \,|\, n \rangle = \sum_n c_n \delta_{nm} = c_m \qquad (2.30)$$

This vector idea is also useful in thinking about how to couple the angular momenta of atoms and molecules. We will see more of this in chapter 3.

2.4.1 Vector Products and Determinants

There are two ways to calculate products of vectors. The scalar, or *dot* product of two vectors, $\mathbf{F}$ and $\mathbf{G}$ is,

$$\mathbf{F} \cdot \mathbf{G} = |\mathbf{F}||\mathbf{G}| \cos \theta \qquad (2.31)$$

where θ is the angle between the two vectors. The dot product can also be written in terms of the vector components:

$$\mathbf{F} = F_x \mathbf{a}_x + F_y \mathbf{a}_y + F_z \mathbf{a}_z \qquad (2.32)$$
$$\mathbf{G} = G_x \mathbf{a}_x + G_y \mathbf{a}_y + G_z \mathbf{a}_z \qquad (2.33)$$
$$\mathbf{F} \cdot \mathbf{G} = F_x G_x + F_y G_y + F_z G_z \qquad (2.34)$$

One useful example of a dot product in physical chemistry is the Zeeman effect (chapter 3), where the magnetic moment $(\hat{\mu}_M)$ of an atom or molecule interacts with a magnetic field, B:

$$\hat{H}^{Zee} = -\hat{\mu}_M \cdot B \qquad (2.35)$$

The other method for calculating the product of two vectors is called the vector, or *cross* product, defined as:

$$\mathbf{F} \times \mathbf{G} = |\mathbf{F}||\mathbf{G}|\mathbf{S} \sin \theta \qquad (2.36)$$

where $\mathbf{S}$ is a unit vector that is perpendicular to the plane formed by $\mathbf{F}$ and $\mathbf{G}$, and θ is the angle between $\mathbf{F}$ and $\mathbf{G}$. The cross product is relevant to classical angular momentum (chapter 3).

The cross product can be written in terms of the vector components as,

$$\mathbf{F} \times \mathbf{G} = (F_y G_z - F_z G_y)\mathbf{a}_x + (F_z G_x - F_x G_z)\mathbf{a}_y + (F_x G_y - F_y G_x)\mathbf{a}_z \tag{2.37}$$

or equivalently as,

$$\mathbf{F} \times \mathbf{G} = \begin{vmatrix} \mathbf{a}_x & \mathbf{a}_y & \mathbf{a}_z \\ F_x & F_y & F_z \\ G_x & G_y & G_z \end{vmatrix} \tag{2.38}$$

This is simply the expansion of a 3x3 determinant, where in general,

$$\begin{vmatrix} k_{11} & k_{12} & k_{13} \\ k_{21} & k_{22} & k_{23} \\ k_{31} & k_{32} & k_{33} \end{vmatrix} \tag{2.39}$$

$$= k_{11}k_{22}k_{33} + k_{21}k_{32}k_{13} + k_{12}k_{23}k_{31} - k_{31}k_{22}k_{13} - k_{21}k_{12}k_{33} - k_{11}k_{23}k_{32}$$

2.5 Matrices

2.5.1 What is a Matrix?

A matrix, M, is just a 2D array of numbers that obeys a set of algebraic rules.

$$M = \begin{pmatrix} M_{11} & M_{12} \\ M_{21} & M_{22} \end{pmatrix} \tag{2.40}$$

The numbers inside the matrix are called *matrix elements*, and they are written generally as M_{ij}, where i is a row and j is a column of the matrix. Matrices can be added and subtracted only if they have the same dimensions:

$$\begin{pmatrix} A & B \\ C & D \end{pmatrix} + \begin{pmatrix} E & F \\ G & H \end{pmatrix} = \begin{pmatrix} A+E & B+F \\ C+G & D+H \end{pmatrix} \tag{2.41}$$

If you multiply any matrix, M, by a scalar quantity, s, each matrix element must be multiplied by that scalar:

$$sM = \begin{pmatrix} sM_{11} & sM_{12} \\ sM_{21} & sM_{22} \end{pmatrix} \tag{2.42}$$

If L, M and N are matrices, the mathematical formula for matrix multiplication is as follows:

$$N = LM \tag{2.43}$$

$$N_{ij} = \sum_k L_{ik} M_{kj} \tag{2.44}$$

Matrix multiplication is not necessarily commutative.

The unit matrix has all matrix elements equal to 0, except for the matrix elements along the diagonal, which are all equal to 1. Any matrix with all matrix elements equal to zero except for those along the diagonal is called a *diagonal* matrix.

$$\mathbf{1} = \begin{pmatrix} 1 & 0 & \cdots & 0 \\ 0 & 1 & \cdots & 0 \\ \vdots & \vdots & \ddots & \vdots \\ 0 & 0 & \cdots & 1 \end{pmatrix} \tag{2.45}$$

The unit matrix is also called the identity matrix, since

$$\mathbf{1}M = M\mathbf{1} \tag{2.46}$$

A matrix, M, may have an inverse, M^{-1}, such that:

$$MM^{-1} = M^{-1}M = \mathbf{1} \tag{2.47}$$

Any good math textbook [68] will help you reach a deeper understanding of the mathematics of matrices and determinants. Computer math software [80] can perform operations such as matrix addition, multiplication, inversion very rapidly.

2.5.2 General Matrix Diagonalization

Matrix diagonalization is the most important mathematical procedure in quantum mechanics. Computer software can perform matrix diagonalization much faster than any human being, but it is useful to see how a simple 2×2 matrix can be diagonalized by hand. Let's work through a few different techniques to see how useful matrix diagonalization can be.

Eigenvalue and Secular Equations

Take a look back at the eigenvalue equation (2.3). This equation can be transformed into a matrix eigenvalue equation,

$$\hat{O}\Psi_i = a_i \Psi_i \tag{2.48}$$

where the operator $\hat{O}$ can be represented by a matrix. As a simple example, let's say that the matrix representation for $\hat{O}$ is:

$$O = \begin{pmatrix} O_{11} & O_{12} \\ O_{21} & O_{22} \end{pmatrix} = \begin{pmatrix} 3 & -2\sqrt{3} \\ -2\sqrt{3} & 2 \end{pmatrix} \tag{2.49}$$

In quantum mechanics, all matrices are *Hermitian*. This means that the matrix representation for the operator is *symmetric* ($O_{ij} = O_{ji}$), *square* (with dimensions $n \times n$) and the matrix elements are all real numbers.

In equation (2.48), the eigenfunction Ψ is represented by a set of column vectors Ψ_i called *eigenvectors*, where i runs from 1 to 2 for a 2×2 matrix. The two eigenvectors are,

$$\Psi_i = \begin{pmatrix} \psi_{1i} \\ \psi_{2i} \end{pmatrix} \tag{2.50}$$

where

$$\Psi = (\Psi_1 \Psi_2) = \begin{pmatrix} \psi_{11} & \psi_{12} \\ \psi_{21} & \psi_{22} \end{pmatrix} \tag{2.51}$$

The matrix representation for a_i is diagonal:

$$a_i = \begin{pmatrix} a_1 & 0 \\ 0 & a_2 \end{pmatrix} \tag{2.52}$$

Our goal is to find the eigenvalues, a_i, and eigenvectors, Ψ_i, that belong to matrix $\hat{O}$. We can rewrite equation (2.48) as,

$$\hat{O}\Psi_i - a_i \Psi_i = 0 \tag{2.53}$$

$$(\hat{O} - a_i \mathbf{1})\Psi_i = 0$$

which is the same thing as two simultaneous equations:

$$(3 - a_1)\psi_{11} - 2\sqrt{3}\psi_{21} = 0 \tag{2.54}$$

$$-2\sqrt{3}\psi_{12} + (2 - a_2)\psi_{22} = 0$$

31

A solution can be found by solving the *secular determinant*:

$$|\hat{O} - a_i \mathbf{1}| = 0 \tag{2.55}$$

In solving the secular determinant, we have,

$$|\hat{O} - a_i \mathbf{1}| = \begin{vmatrix} 3 - a_i & -2\sqrt{3} \\ -2\sqrt{3} & 2 - a_i \end{vmatrix} = (3 - a_i)(2 - a_i) - 12 = 0 \tag{2.56}$$

This becomes a quadratic equation,

$$a_i^2 - 5a_i - 6 = 0 \tag{2.57}$$

with roots (eigenvalues):

$$a_1 = 6, a_2 = -1 \tag{2.58}$$

$$a_i = \begin{pmatrix} 6 & 0 \\ 0 & -1 \end{pmatrix} \tag{2.59}$$

In order to find the eigenvectors, we can plug the eigenvalues back into equation (2.54):

$$(-3)\psi_{11} - 2\sqrt{3}\psi_{21} = 0 \tag{2.60}$$

$$-2\sqrt{3}\psi_{12} + (3)\psi_{22} = 0$$

Since the eigenvectors must be orthonormal,

$$\psi_{1i}^2 + \psi_{2i}^2 = 1$$

$$\psi_{11}^2 + \psi_{21}^2 = 1 \tag{2.61}$$

$$\psi_{12}^2 + \psi_{22}^2 = 1$$

we can solve to obtain the eigenvector matrix:

$$\Psi = (\Psi_1 \Psi_2) = \begin{pmatrix} -\frac{2}{\sqrt{7}} & \sqrt{\frac{3}{7}} \\ \sqrt{\frac{3}{7}} & \frac{2}{\sqrt{7}} \end{pmatrix} \tag{2.62}$$

The Similarity Transformation

The eigenvectors that we obtained are very useful because they can also be used to diagonalize the matrix to obtain corresponding eigenvalues. The following equation is called a *similarity transformation*:

$$\Psi^{-1}\hat{O}\Psi = a \tag{2.63}$$

The $\hat{O}$ matrix is diagonalized by the eigenvector matrix:

$$\begin{pmatrix} -\frac{2}{\sqrt{7}} & \sqrt{\frac{3}{7}} \\ \sqrt{\frac{3}{7}} & \frac{2}{\sqrt{7}} \end{pmatrix} \begin{pmatrix} 3 & -2\sqrt{3} \\ -2\sqrt{3} & 2 \end{pmatrix} \begin{pmatrix} -\frac{2}{\sqrt{7}} & \sqrt{\frac{3}{7}} \\ \sqrt{\frac{3}{7}} & \frac{2}{\sqrt{7}} \end{pmatrix} = \begin{pmatrix} 6 & 0 \\ 0 & -1 \end{pmatrix} \tag{2.64}$$

2.5.3   Matrix Diagonalization in Quantum Mechanics

The simple 2×2 matrix example that we have been studying can be used to analyze interactions, or *perturbations*, between two quantized energy levels in an atom or molecule. Let's start with an example matrix representation model for a Hamiltonian operator. If the matrix that we construct is already diagonal, the nonzero matrix elements give us two energy levels (eigenvalues):

$$H = \begin{pmatrix} H_{11} & 0 \\ 0 & H_{22} \end{pmatrix} = \begin{pmatrix} E_1 & 0 \\ 0 & E_2 \end{pmatrix} \tag{2.65}$$

If the model Hamiltonian is the true, exact Hamiltonian for the system, then the problem has been solved. However, we may not know ahead of time what the exact Hamiltonian is for the system, and we might have to take an educated guess for the matrix representation. What happens if the model Hamiltonian that we pick is not exactly the true Hamiltonian for the system? If the model Hamiltonian is not the true Hamiltonian, there must be some extra interaction, a perturbation, that we left out of our Hamiltonian model. This neglected part of the Hamiltonian gives rise to non-zero off-diagonal matrix elements, P.

$$H = \begin{pmatrix} H_{11} & H_{12} \\ H_{21} & H_{22} \end{pmatrix} = \begin{pmatrix} E_1 & P \\ P & E_2 \end{pmatrix} \tag{2.66}$$

33

Let's say you had a system with two states (figure 2.5). This could be an electron in a magnetic field with two orientations, spin up and spin down. The P perturbation mixes the character of the two wavefunctions belonging to the two states, and shifts the energies of the levels in opposite directions. The degree of mixing and shifting depends on magnitudes of P and $\Delta E = E_2 - E_1$.

Simple Solution for a 2×2 Matrix

In the special case of a 2×2 matrix, there is a simple formula for obtaining the eigenvalues. We can solve the problem exactly using the secular equation.

$$H = \begin{pmatrix} E_1 & P \\ P & E_2 \end{pmatrix} \tag{2.67}$$

$$(E_1 - E)(E_2 - E) - P^2 = 0 \tag{2.68}$$

The eigenvalues are,

$$E_\pm = \frac{E_1 + E_2}{2} \pm \frac{\sqrt{(E_1 - E_2)^2 + 4P^2}}{2} \tag{2.69}$$

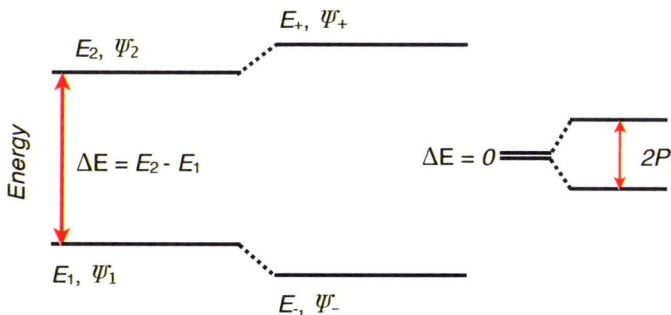

Figure 2.5: Effect of a perturbation applied to a 2-level system. If $E_1 = E_2 = E$, then $E_\pm = \frac{E_1+E_2}{2} \pm \frac{\sqrt{(E_1-E_2)^2+4P^2}}{2}$ becomes $E_\pm = E \pm P$.

Rotation Matrices

A rotation matrix, R, can also be used to diagonalize a 2×2 Hamiltonian matrix using the similarity transformation,

$$R^{-1}\hat{H}R = E \tag{2.70}$$

where

$$R = \begin{pmatrix} \cos\theta & -\sin\theta \\ \sin\theta & \cos\theta \end{pmatrix} \tag{2.71}$$

and

$$H = \begin{pmatrix} E_1 & P \\ P & E_2 \end{pmatrix} \tag{2.72}$$

The R matrix simply represents the rotation of a vector through an angle, θ. The angle θ is chosen to make the matrix diagonal.

$$\tan 2\theta = \frac{2P}{E_2 - E_1} \tag{2.73}$$

$$\theta = \frac{1}{2}\arctan\frac{2P}{E_2 - E_1} \tag{2.74}$$

Rotation matrices are very useful because we can use them to write down expressions for the two new mixed wavefunctions:

$$|\Psi_-\rangle = \cos\theta\,|\psi_1\rangle - \sin\theta\,|\psi_2\rangle \tag{2.75}$$

$$|\Psi_+\rangle = \sin\theta\,|\psi_1\rangle + \cos\theta\,|\psi_2\rangle \tag{2.76}$$

2.6 The Hamiltonian Matrix

2.6.1 Matrix Elements

When we write Dirac notation with a Hamiltonian operator sandwiched in between a bra and ket,

$$\left\langle m\,\middle|\,\hat{H}\,\middle|\,n\right\rangle \tag{2.77}$$

we can use the shorthand,

$$H_{mn} = \left\langle m\,\middle|\,\hat{H}\,\middle|\,n\right\rangle \tag{2.78}$$

to describe a matrix element, where m and n are the rows and columns of a matrix. This is called the matrix representation for the Hamiltonian operator, where

$$H_{mn} = \begin{pmatrix} H_{11} & H_{12} & H_{13} & H_{14} \\ H_{21} & H_{22} & H_{23} & H_{24} \\ H_{31} & H_{32} & H_{33} & H_{34} \\ H_{41} & H_{42} & H_{43} & H_{44} \end{pmatrix} \tag{2.79}$$

2.6.2 Diagonalization of the Hamiltonian Matrix

This matrix representation of the Hamiltonian operator can be used to solve the time-independent Schrödinger equation,

$$H\,|\psi\rangle = E\,|\psi\rangle \tag{2.80}$$

From what we have already learned, we can rewrite $|\psi\rangle$ as a linear combination of basis states, $|n\rangle$:

$$H\,|\psi\rangle = H\sum_n c_n\,|n\rangle = \sum_n c_n H\,|n\rangle \tag{2.81}$$

$$E \left| \psi \right\rangle = E \sum_n c_n \left| n \right\rangle$$

We can multiply from the left by one of the basis states as a bra, $\left\langle m \right|$, to get:

$$\sum_n c_n \left\langle m | H | n \right\rangle = E \sum_n c_n \left\langle m | n \right\rangle = E c_m \qquad (2.82)$$

Next, translate this into matrix notation,

$$\sum_n H_{mn} c_n = E c_m \qquad (2.83)$$

where

$$H_{mn} = \begin{pmatrix} H_{11} & H_{12} & H_{13} & H_{14} \\ H_{21} & H_{22} & H_{23} & H_{24} \\ H_{31} & H_{32} & H_{33} & H_{34} \\ H_{41} & H_{42} & H_{43} & H_{44} \end{pmatrix} \qquad (2.84)$$

If we can *diagonalize* this Hamiltonian so that $H_{mn} = 0$ unless $m = n$,

$$H_{mm} c_m = E c_m \qquad (2.85)$$

the equation becomes,

$$H_{mm} = \begin{pmatrix} E_1 & 0 & 0 & 0 \\ 0 & E_2 & 0 & 0 \\ 0 & 0 & E_3 & 0 \\ 0 & 0 & 0 & E_4 \end{pmatrix} \qquad (2.86)$$

$$H_{mm} = E \qquad (2.87)$$

The energies (eigenvalues) are just the matrix elements along the diagonal. The key point here is that diagonalizing the Hamiltonian matrix is equivalent to solving the Schrödinger equation to obtain the energies of an atom or molecule.

2.7   The Harmonic Oscillator

Swinging pendulums, vibrating strings, electromagnetic fields and vibrations between atoms in a molecule can be described as harmonic oscillators. We can use the matrix representations for the position, $\hat{x}$, and momentum, $\hat{p}_x$, operators to calculate the energies of the quantized harmonic oscillator.

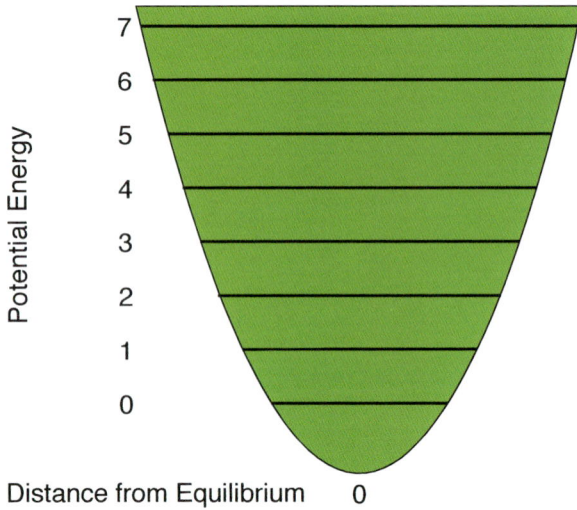

Figure 2.6: The potential energy of a harmonic oscillator with an evenly spaced ladder of energy levels.

2.7.1 The Creation and Annihilation Operators

The energy levels in figure 2.6 are labeled by a quantum number, n. The levels have energies,

$$E_n = \left(n + \frac{1}{2}\right)\hbar\omega \qquad (2.88)$$

where,

$$\omega = \left(\frac{k}{m}\right)^{\frac{1}{2}} \qquad (2.89)$$

k is called the force constant, m is the mass and $n = 0, 1, 2 \ldots$. The matrix elements of the harmonic oscillator are in general,

$$H_{mn} = \langle m|\hat{H}|n\rangle = \left(n + \frac{1}{2}\right)\hbar\omega\delta_{mn} \qquad (2.90)$$

where this equation defines an infinite diagonal matrix, H:

$$H = \begin{pmatrix} \hbar\omega/2 & 0 & \cdots & 0 \\ 0 & 3\hbar\omega/2 & \cdots & 0 \\ \vdots & \vdots & \ddots & \vdots \\ 0 & 0 & \cdots & \left(n+\frac{1}{2}\right)\hbar\omega \end{pmatrix} \qquad (2.91)$$

The rows and columns are labeled by m and n, respectively ($m = 0, 1, 2, \ldots$. $n = 0, 1, 2, \ldots$). If we represent the problem this way, the state $|n\rangle$ is a column vector with an infinite number of rows, and all elements zero except for the nth-row element, which is equal to one. This is the same vector idea that we discussed earlier in the chapter.

$$|n\rangle = \begin{pmatrix} 0 \\ 0 \\ 0 \\ \vdots \\ 0 \\ 1 \\ 0 \\ 0 \\ 0 \\ \vdots \\ \vdots \end{pmatrix} \qquad (2.92)$$

Matrix representations like this are the foundation of the matrix mechanics that Heisenberg, Born and Jordan came up with. Just like our matrix representation of $\hat{H}$ and $|n\rangle$, they came up with matrix representations for $\hat{x}$ and $\hat{p}$. The Heisenberg Uncertainty Principle is expressed as,

$$[\hat{x}, \hat{p}] = i\hbar \mathbf{1} \qquad (2.93)$$

where $\mathbf{1}$ is the unit matrix.

Our goal now is to evaluate the matrix elements for the position and momentum operators,

$$x_{mn} = \langle m|\hat{x}|n\rangle \qquad (2.94)$$

$$p_{mn} = \langle m|\hat{p}|n\rangle \qquad (2.95)$$

and use them to solve the harmonic oscillator Schrödinger equation. We can do this using the creation and annihilation operators [40, 44], defined as:

$$\hat{a} = \sqrt{\frac{m\omega}{\hbar}} \frac{1}{\sqrt{2}} \hat{x} + i\sqrt{\frac{\hbar}{m\omega}} \frac{1}{\hbar\sqrt{2}} \hat{p} \qquad (2.96)$$

$$\hat{a}^\dagger = \sqrt{\frac{m\omega}{\hbar}} \frac{1}{\sqrt{2}} \hat{x} - i\sqrt{\frac{\hbar}{m\omega}} \frac{1}{\hbar\sqrt{2}} \hat{p} \qquad (2.97)$$

These operators have the following eigenvalue equations,

$$\hat{a}^\dagger \ket{n} = \sqrt{n+1} \ket{n+1} \qquad (2.98)$$

$$\hat{a} \ket{n} = \sqrt{n} \ket{n-1} \qquad (2.99)$$

where you can see that the operators applied to a ket, $\ket{n}$, create a quantum of energy or annihilate a quantum of energy. With these equations, we can get expressions for momentum and position in terms of the creation and annihilation operators:

$$\hat{x} = \sqrt{\frac{\hbar}{m\omega}} \frac{1}{\sqrt{2}} (\hat{a}^\dagger + \hat{a}) \qquad (2.100)$$

$$\hat{p} = \sqrt{\frac{m\omega}{\hbar}} \frac{i\hbar}{\sqrt{2}} (\hat{a}^\dagger - \hat{a}) \qquad (2.101)$$

Next, we should be able to write down the matrix elements of a and $a^\dagger$:

$$\bra{m}\hat{a}^\dagger\ket{n} = \sqrt{n+1} \braket{m|n+1} = \sqrt{n+1}\delta_{m,n+1} \qquad (2.102)$$

$$\bra{m}\hat{a}\ket{n} = \sqrt{n} \braket{m|n-1} = \sqrt{n}\delta_{m,n-1} \qquad (2.103)$$

We should now be able to use these results to get the matrix elements of x and p:

$$\bra{m}\hat{x}\ket{n} = \sqrt{\frac{\hbar}{m\omega}} \frac{1}{\sqrt{2}} (\sqrt{n+1}\delta_{m,n+1} + \sqrt{n}\delta_{m,n-1}) \qquad (2.104)$$

$$\bra{m}\hat{p}\ket{n} = \sqrt{\frac{m\omega}{\hbar}} \frac{i\hbar}{\sqrt{2}} (\sqrt{n+1}\delta_{m,n+1} - \sqrt{n}\delta_{m,n-1}) \qquad (2.105)$$

You can tell by the Kronecker deltas that each of the four above matrices for x, p, a and $a^\dagger$ has only off-diagonal elements, lying just above and/or below the main diagonal. We can explicitly write out the matrices for a and $a^\dagger$:

$$a^\dagger = \begin{pmatrix} 0 & 0 & 0 & 0 & 0 & 0 \\ \sqrt{1} & 0 & 0 & 0 & 0 & 0 \\ 0 & \sqrt{2} & 0 & 0 & 0 & 0 \\ 0 & 0 & \sqrt{3} & 0 & 0 & 0 \\ 0 & 0 & 0 & \sqrt{4} & 0 & 0 \\ 0 & 0 & 0 & 0 & \sqrt{5} & 0 \end{pmatrix} \quad (2.106)$$

Of course, these are both infinite matrices that go on forever.

$$a = \begin{pmatrix} 0 & \sqrt{1} & 0 & 0 & 0 & 0 \\ 0 & 0 & \sqrt{2} & 0 & 0 & 0 \\ 0 & 0 & 0 & \sqrt{3} & 0 & 0 \\ 0 & 0 & 0 & 0 & \sqrt{4} & 0 \\ 0 & 0 & 0 & 0 & 0 & \sqrt{5} \\ 0 & 0 & 0 & 0 & 0 & 0 \end{pmatrix} \quad (2.107)$$

We can also write out the matrices for x and p:

$$x = \sqrt{\frac{\hbar}{2m\omega}} \begin{pmatrix} 0 & \sqrt{1} & 0 & 0 & 0 & 0 \\ \sqrt{1} & 0 & \sqrt{2} & 0 & 0 & 0 \\ 0 & \sqrt{2} & 0 & \sqrt{3} & 0 & 0 \\ 0 & 0 & \sqrt{3} & 0 & \sqrt{4} & 0 \\ 0 & 0 & 0 & \sqrt{4} & 0 & \sqrt{5} \\ 0 & 0 & 0 & 0 & \sqrt{5} & 0 \end{pmatrix} \quad (2.108)$$

$$p = \sqrt{\frac{\hbar m\omega}{2}} \begin{pmatrix} 0 & -i\sqrt{1} & 0 & 0 & 0 & 0 \\ i\sqrt{1} & 0 & -i\sqrt{2} & 0 & 0 & 0 \\ 0 & i\sqrt{2} & 0 & -i\sqrt{3} & 0 & 0 \\ 0 & 0 & i\sqrt{3} & 0 & -i\sqrt{4} & 0 \\ 0 & 0 & 0 & i\sqrt{4} & 0 & -i\sqrt{5} \\ 0 & 0 & 0 & 0 & i\sqrt{5} & 0 \end{pmatrix} \quad (2.109)$$

These matrices satisfy the commutation relationship, $[x, p] = i\hbar \mathbf{1}$:

$$[x, p] = \begin{pmatrix} i\hbar & 0 & 0 & 0 & 0 & 0 \\ 0 & i\hbar & 0 & 0 & 0 & 0 \\ 0 & 0 & i\hbar & 0 & 0 & 0 \\ 0 & 0 & 0 & i\hbar & 0 & 0 \\ 0 & 0 & 0 & 0 & i\hbar & 0 \\ 0 & 0 & 0 & 0 & 0 & i\hbar \end{pmatrix} \quad (2.110)$$

2.7.2 Solution to the Harmonic Oscillator Problem

Finally, we can solve the harmonic oscillator Schrödinger equation using the x and p matrix representations. The harmonic oscillator Hamiltonian matrix is $\hat{H} = \frac{p^2}{2m} + \frac{1}{2}kx^2$. Plugging in our matrices, we obtain,

$$H^{H.O.} = \begin{pmatrix} \frac{1}{2}\hbar\omega & 0 & 0 & 0 & 0 & 0 \\ 0 & \frac{3}{2}\hbar\omega & 0 & 0 & 0 & 0 \\ 0 & 0 & \frac{5}{2}\hbar\omega & 0 & 0 & 0 \\ 0 & 0 & 0 & \frac{7}{2}\hbar\omega & 0 & 0 \\ 0 & 0 & 0 & 0 & \frac{9}{2}\hbar\omega & 0 \\ 0 & 0 & 0 & 0 & 0 & \frac{11}{2}\hbar\omega \end{pmatrix} \qquad (2.111)$$

The eigenvalues are exactly what we expected.

2.8 Statistical Mechanics in Chemistry

When we learn chemistry in the classroom, it is traditionally divided into two main groups. On one hand, we learn about the macroscopic world of chemicals, dealing with bulk properties of matter. This world includes thermodynamics (where concepts like enthalpy, entropy and free energy are introduced) and chemical kinetics (for understanding rates of reactions). On the other hand is the microscopic world, where properties of individual atoms and molecules are key. This world is governed by quantum mechanics.

When mixing chemicals in a laboratory, or measuring ingredients in the kitchen, we can easily use stoichiometry to translate masses from the familiar macroscopic world to the numbers and ratios of atoms and molecules in the microscopic world using concepts like moles and Avogadro's constant (6.02×10^{23} mol^{-1}). We know that the properties of bulk matter ultimately depend on properties of individual atoms and molecules, but what is the mathematical connection between a large ensemble of atoms and molecules that gives rise to bulk properties like temperature, pressure, heat capacity, or the distribution of blackbody radiation?

If you want to understand macroscopic chemistry in terms of the properties of individual molecules, you need statistics and probability. Statistical mechanics is the bridge between the macroscopic and microscopic worlds. We have seen Einstein's derivation of Planck's quantum blackbody radiation law in chapter 1. Next, we will examine an alternative derivation using statistical mechanics.

2.8.1   Planck's Blackbody Law Revisited

In 1900, Rayleigh derived the classical mechanics equation for the energy density inside a blackbody cavity, based on the equipartition law of energy. He made the fundamental assumption that the average energy for every degree of freedom of the oscillators that make up the walls of the cavity is $k_B T$. The classical energy density, $\rho_\nu(T)d\nu$, of blackbody radiation in the range ν to $\nu + d\nu$ is,

$$d\rho(\nu, T) = \rho_\nu(T)d\nu = \frac{\langle E \rangle \, N(\nu)}{V} d\nu \qquad (2.112)$$

where $\langle E \rangle = k_B T$ is the average energy of the oscillators that are in equilibrium with the radiation field in the cavity. $N(\nu)$ is the number of modes of the electromagnetic radiation field in the cavity with volume V. Describing the number of cavity modes (degrees of freedom) for standing waves of electromagnetic radiation as,

$$\frac{N(\nu)}{V} = \frac{2 \times 4\pi\nu^2}{c^3} \qquad (2.113)$$

the resulting Rayleigh-Jeans law is,

$$\rho_\nu(T)d\nu = \frac{8\pi\nu^2 d\nu}{c^3} k_B T \qquad (2.114)$$

This classical mechanics expression failed to predict the intensity of blackbody emission at all frequencies, because it doesn't account for the quantum nature of atoms and molecules.

How can one go from Rayleigh's classical mechanics equation to Planck's quantum formula for blackbody radiation? The energy of each mode of the electromagnetic field is not continuous (as in the classical average value, $k_B T$), but is an integral multiple of $\epsilon = h\nu$. A fundamental premise of statistical mechanics says that the probability, P_n, of finding the system in the nth quantum state with energy $E_n = nh\nu$ is equal to the Boltzmann factor divided by the partition function,

$$P_n = \frac{e^{\frac{-E_n}{k_B T}}}{\sum_n e^{\frac{-E_n}{k_B T}}} \qquad (2.115)$$

43

assuming thermal equilibrium, so that there is a well-defined temperature. The average energy of the system is,

$$\langle E(\nu) \rangle = \sum_{n=0}^{\infty} E_n P_n = \frac{\sum_{n=0}^{\infty} n\epsilon e^{\frac{-n\epsilon}{k_B T}}}{\sum_{n=0}^{\infty} e^{\frac{-n\epsilon}{k_B T}}} = \frac{-\epsilon \frac{d}{dx} \sum_{n=0}^{\infty} e^{-nx}}{\sum_{n=0}^{\infty} e^{-nx}} \qquad (2.116)$$

where $x = \epsilon/k_B T$. The sums can be replaced by their geometric series limit, $(1 - e^{-x})^{-1}$, to give,

$$\langle E(\nu) \rangle = \frac{h\nu}{e^{\frac{h\nu}{k_B T}} - 1} \qquad (2.117)$$

consistent with Bose-Einstein statistics for a system of indistinguishable bosons. Replacing the classical average energy from equation (2.114) with the quantum average energy in equation (2.117) gives Planck's quantum blackbody radiation law!

$$d\rho(\nu, T) = \rho_\nu(T)d\nu = \frac{8\pi\nu^2}{c^3} \frac{h\nu}{e^{\frac{h\nu}{k_B T}} - 1} d\nu \qquad (2.118)$$

Chapter 3

Angular Momentum

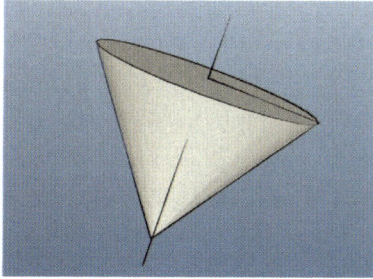

Figure 3.1: A precessing spinning top would trace out a cone shape.

Children used to play with spinning tops. If you have seen one of these tops, you know that after the top spins for a while, gravity will make the top fall. However, the top doesn't simply topple over onto the ground. Instead, the torque from gravitational pull on the top causes the principal axis of the top to swing around and trace out the surface of a cone. This motion along a cone is called precession. This classical picture of angular momentum is useful in understanding the purely mathematical quantum angular momentum. There are many great texts on angular momenta in quantum mechanics [9, 10, 22, 65, 84, 97]. The mathematics of quantum angular momentum is fascinating because there are so many wonderful mathematical relationships that one can derive to describe physical systems.

In this chapter, we will examine the main results of angular momentum matrix element calculations using angular momentum operators and basis functions. We will then apply these results to solve real problems involving atoms that will help you make a connection to modern technology based on quantum mechanics.

3.1 Angular Momenta in Physical Chemistry

Quantum angular momentum is key to understanding atomic and molecular structure. We will develop quantum angular momentum concepts to see how the coupling of angular momenta of electrons and nuclei affect the energy level structure and properties of atoms and molecules.

There are five main goals that we will reach in this chapter:
(1) Write down relevant angular momentum operators and eigenstates.
(2) Calculate angular momentum eigenvalue equations and matrix elements.
(3) Mathematically represent the coupling of angular momenta using $3j$-Symbols.
(4) Build a Hamiltonian matrix for the coupling of two angular momenta.
(5) Build a Hamiltonian matrix for an atom in an external field.

3.2 Angular Momentum Operators

3.2.1 Some Basic Definitions

Classical Mechanics

Any particle can be described as having a certain mass, m, and velocity, $\mathbf{v}$, at a certain position, $\mathbf{r}$, from an origin point. According to classical mechanics, you already know that such a particle will have a linear momentum, $\mathbf{p}$, defined as,

$$\mathbf{p} = m\mathbf{v} \tag{3.1}$$

and angular momentum, $\mathbf{l}$,

$$\mathbf{l} = \mathbf{r} \times \mathbf{p} \tag{3.2}$$

$$\mathbf{l} = \mathbf{r} \times \mathbf{p} = \begin{vmatrix} \mathbf{a}_x & \mathbf{a}_y & \mathbf{a}_z \\ x & y & z \\ p_x & p_y & p_z \end{vmatrix} \tag{3.3}$$

Figure 3.2 shows the cross product of $\mathbf{r}$ and $\mathbf{p}$ to give $\mathbf{l}$.

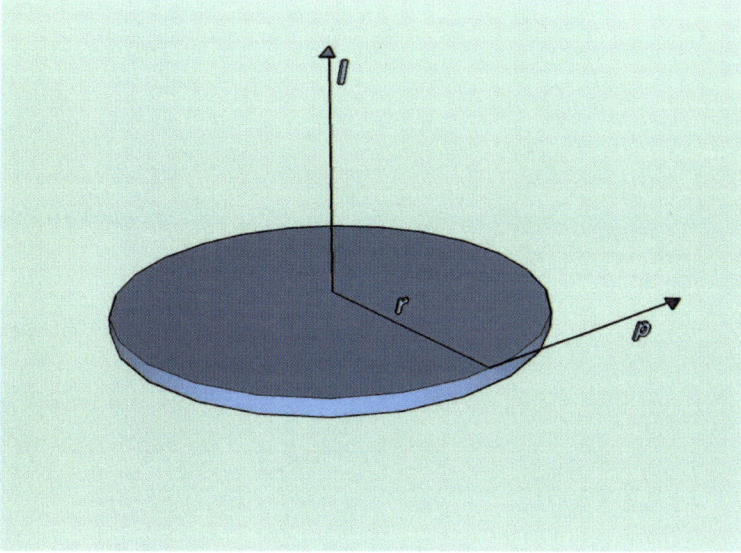

Figure 3.2: Orbital angular momentum.

Quantum Mechanics

Let us look at the mathematical form of momentum operators in quantum mechanics. Linear momentum along x, y and z can be represented by differentiation with respect to x, y and z. The Cartesian components of $\hat{p}$ are:

$$\hat{p}_x = -i\hbar \frac{\partial}{\partial x}, \quad \hat{p}_y = -i\hbar \frac{\partial}{\partial y}, \quad \hat{p}_z = -i\hbar \frac{\partial}{\partial z} \tag{3.4}$$

From now on, we will use a system of units where $\hbar = 1$ and write,

$$\hat{p}_x = -i \frac{\partial}{\partial x}, \quad \hat{p}_y = -i \frac{\partial}{\partial y}, \quad \hat{p}_z = -i \frac{\partial}{\partial z} \tag{3.5}$$

Classical angular momentum can also be transformed into a quantum mechanical operator called orbital angular momentum.

The Cartesian components of $\hat{l}$ are:

$$\hat{l}_x = y\hat{p}_z - z\hat{p}_y = -i\left(y\frac{\partial}{\partial z} - z\frac{\partial}{\partial y}\right) \qquad (3.6)$$

$$\hat{l}_y = z\hat{p}_x - x\hat{p}_z = -i\left(z\frac{\partial}{\partial x} - x\frac{\partial}{\partial z}\right) \qquad (3.7)$$

$$\hat{l}_z = x\hat{p}_y - y\hat{p}_x = -i\left(x\frac{\partial}{\partial y} - y\frac{\partial}{\partial x}\right) \qquad (3.8)$$

3.2.2 Commutation Relations

We have already seen the general form of a commutator,

$$[\hat{A}, \hat{B}] = \hat{A}\hat{B} - \hat{B}\hat{A} \qquad (3.9)$$

If the observables that correspond to operators $\hat{A}$ and $\hat{B}$ are to be simultaneously measurable to any arbitrary precision, the operators must commute,

$$[\hat{A}, \hat{B}] = 0 \qquad (3.10)$$

If they are not simultaneously measurable to any precision, then,

$$[\hat{A}, \hat{B}] \neq 0 \qquad (3.11)$$

Here are the commutation relationships for position and linear momentum:

$$[\hat{x}, \hat{p}_x] = i, \ [\hat{x}, \hat{p}_y] = 0, \ [\hat{x}, \hat{p}_z] = 0 \qquad (3.12)$$

Here are the commutation relationships for orbital angular momentum, **l**:

$$[\hat{l}_x, \hat{l}_y] = i\hat{l}_z, \ [\hat{l}_y, \hat{l}_z] = i\hat{l}_x, \ [\hat{l}_z, \hat{l}_x] = i\hat{l}_y \qquad (3.13)$$

3.2.3 General Angular Momentum Operators

If we define a general quantum angular momentum, **j**, the commutation relationships are:

$$[\hat{j}_x, \hat{j}_y] = i\hat{j}_z, \ [\hat{j}_y, \hat{j}_z] = i\hat{j}_x, \ [\hat{j}_z, \hat{j}_x] = i\hat{j}_y \qquad (3.14)$$

The magnitude of this general angular momentum is related to its components in the same way you would calculate the magnitude of a vector:

$$\hat{j}^2 = \hat{j}_x^2 + \hat{j}_y^2 + \hat{j}_z^2 \qquad (3.15)$$

The relevant commutation relationships are:

$$[\hat{j}^2, \hat{j}_x] = [\hat{j}^2, \hat{j}_y] = [\hat{j}^2, \hat{j}_z] = 0 \qquad (3.16)$$

All quantum angular momentum operators behave according to equations (3.15) and (3.16).

Ladder (Shift) Operators

Ladder operators are linear combinations of angular momentum operators. They are most useful because they can be used to evaluate matrix elements of angular momentum operators.

The *raising* operator is $\hat{j}_+$:

$$\hat{j}_+ = \hat{j}_x + i\hat{j}_y \qquad (3.17)$$

The *lowering* operator is $\hat{j}_-$:

$$\hat{j}_- = \hat{j}_x - i\hat{j}_y \qquad (3.18)$$

The relevant commutation relationships are:

$$[\hat{j}^2, \hat{j}_\pm] = 0 \qquad (3.19)$$

$$[\hat{j}_z, \hat{j}_\pm] = \pm\hat{j}_\pm \qquad (3.20)$$

$$[\hat{j}_+, \hat{j}_-] = 2\hat{j}_z \qquad (3.21)$$

3.3 Angular Momentum Eigenvalue Equations

In order to construct angular momentum eigenvalue equations, we must first write down what types of angular momentum eigenstates we will encounter. The letters in the kets are the quantum numbers that describe the states of the system. Each ket contains a quantum number and its projection.

- General angular momentum: $|j, m_j\rangle$

- Orbital angular momentum: $|l, m_l\rangle$

- Electronic spin: $|s, m_s\rangle$

- Nuclear spin: $|i, m_i\rangle$

- Hyperfine: $|f, m_f\rangle$

For any given quantum number, j, there are $2j + 1$ different possible values for the projection quantum number, m_j, where $m_j = j, \ j - 1, \ldots, -j$. The relevant eigenvalue equations are:

$$\hat{j}^2 |j, m_j\rangle = j(j + 1) |j, m_j\rangle \tag{3.22}$$

$$\hat{j}_z |j, m_j\rangle = m_j |j, m_j\rangle \tag{3.23}$$

where the operators perform an operation on a general ket to give a constant (in terms of quantum numbers j and m_j) multiplied by the same general ket. Take note of the difference between an operator with a hat $(\hat{j}^2)$ and a quantum number with no hat (j). The raising and lowering operators perform similarly,

$$\hat{j}_+ |j, m_j\rangle = [j(j + 1) - m_j(m_j + 1)]^{1/2} |j, m_j + 1\rangle \tag{3.24}$$

$$\hat{j}_- |j, m_j\rangle = [j(j + 1) - m_j(m_j - 1)]^{1/2} |j, m_j - 1\rangle \tag{3.25}$$

where you can see that the $\hat{j}_\pm$ are raising/lowering operators because they generate a state with the same magnitude, but projection quantum number one unit greater/smaller.

3.4 Angular Momentum Matrix Elements

The matrix elements of the angular momentum operators are:

$$\langle j, m_j | \hat{j}^2 | j', m_j' \rangle = j(j+1)\delta_{j,j'}\delta_{m_j,m_j'} \qquad (3.26)$$

$$\langle j, m_j | \hat{j}_z | j', m_j' \rangle = m_j\delta_{j,j'}\delta_{m_j,m_j'} \qquad (3.27)$$

$$\langle j, m_j | \hat{j}_\pm | j', m_j' \rangle = \left[j(j+1) - m_j'(m_j' \pm 1)\right]^{1/2}\delta_{j,j'}\delta_{m_j,m_j'\pm 1} \qquad (3.28)$$

3.5 Angular Momentum Coupling

3.5.1 The Coupled vs. Uncoupled Pictures

Let's say there are two sources of angular momentum, j_1 and j_2, that describe a chemical system (atom or molecule). The system might be an atom with both spin and orbital angular momentum. For example, it could be a p-electron ($l = 1$ and $s = 1/2$). We want to describe the whole system mathematically, but there are two ways of doing so. Fortunately, both ways involve vector addition.

One mathematical description is called the *uncoupled* picture:

$$|j_1 m_{j1}; j_2 m_{j2}\rangle \qquad (3.29)$$

where j_1, m_{j1}, j_2 and m_{j2} are all *good* quantum numbers for the system.

The other description is called the *coupled* picture:

$$|j_1 j_2; j m_j\rangle \qquad (3.30)$$

where j_1 and j_2 are coupled together to give a new angular momentum, j, and its projection, m_j. m_{j1} and m_{j2} are no longer good quantum numbers for the system under the coupled picture.

In order to understand this further, we must write out the mathematical rule for specifying $j = j_1 + j_2$ and $m_j = m_{j1} + m_{j2}$.

3.5.2 The Clebsch-Gordan Series

The Clebsch-Gordan Series tells us what possible values of j can be obtained from the coupling of j_1 and j_2:

$$j = j_1 + j_2, \quad j_1 + j_2 - 1, \ldots, |j_1 - j_2| \tag{3.31}$$

For example, let's say that we are studying an atom with two angular momenta that are coupled together, $j_1 = \frac{1}{2}$ and $j_2 = \frac{3}{2}$. The first thing we can do is calculate the maximum and minimum values for j: $\frac{1}{2} + \frac{3}{2} = 2$ and $|\frac{1}{2} - \frac{3}{2}| = 1$. The degeneracy of the system is $(2j_1 + 1)(2j_2 + 1) = 2 \times 4 = 8$ total states that we can write as kets. The kets for the uncoupled representation are:

$$
\begin{array}{cc}
\left|\frac{1}{2}, \frac{1}{2}; \frac{3}{2}, +\frac{3}{2}\right\rangle & \left|\frac{1}{2}, -\frac{1}{2}; \frac{3}{2}, +\frac{3}{2}\right\rangle \\[6pt]
\left|\frac{1}{2}, \frac{1}{2}; \frac{3}{2}, +\frac{1}{2}\right\rangle & \left|\frac{1}{2}, -\frac{1}{2}; \frac{3}{2}, +\frac{1}{2}\right\rangle \\[6pt]
\left|\frac{1}{2}, \frac{1}{2}; \frac{3}{2}, -\frac{1}{2}\right\rangle & \left|\frac{1}{2}, -\frac{1}{2}; \frac{3}{2}, -\frac{1}{2}\right\rangle \\[6pt]
\left|\frac{1}{2}, \frac{1}{2}; \frac{3}{2}, -\frac{3}{2}\right\rangle & \left|\frac{1}{2}, -\frac{1}{2}; \frac{3}{2}, -\frac{3}{2}\right\rangle
\end{array}
\tag{3.32}
$$

The kets for the coupled representation are:

$$
\left|\frac{1}{2}, \frac{3}{2}; 2, +2\right\rangle \quad \left|\frac{1}{2}, \frac{3}{2}; 2, +1\right\rangle \quad \left|\frac{1}{2}, \frac{3}{2}; 2, 0\right\rangle \quad \left|\frac{1}{2}, \frac{3}{2}; 2, -1\right\rangle \quad \left|\frac{1}{2}, \frac{3}{2}; 2, -2\right\rangle
\tag{3.33}
$$

$$
\left|\frac{1}{2}, \frac{3}{2}; 1, +1\right\rangle \quad \left|\frac{1}{2}, \frac{3}{2}; 1, 0\right\rangle \quad \left|\frac{1}{2}, \frac{3}{2}; 1, -1\right\rangle
$$

3.5.3 Wigner 3j-Symbols and Clebsch-Gordan Coefficients

There is an important mathematical relationship between the coupled and uncoupled pictures:

$$|j_1 j_2; j m_j\rangle = \sum_{m_{j1}, m_{j2}} W3j(j_1, j_2, j; m_{j1}, m_{j2}, m_j) |j_1 m_{j1}; j_2 m_{j2}\rangle \tag{3.34}$$

where the $W3j(j_1, j_2, j; m_{j1}, m_{j2}, m_j)$ are called Wigner 3j-Symbols, or Clebsch-Gordan coefficients. These are simply numbers that can be calculated using math software [80]. We can also write equation (3.34) in a simpler form by eliminating the quantum numbers that never change from one ket to the next:

$$|j, m_j\rangle = \sum_{m_{j1}, m_{j2}} W3j(j_1, j_2, j; m_{j1}, m_{j2}, m_j)\, |m_{j1}; m_{j2}\rangle \quad (3.35)$$

where

$$W3j(j_1, j_2, j; m_{j1}, m_{j2}, m_j) =$$
$$(-1)^{j_1 - j_2 + m_j} \sqrt{2j+1} \begin{pmatrix} j_1 & j_2 & j \\ m_{j1} & m_{j2} & -m_j \end{pmatrix} \quad (3.36)$$

The $\begin{pmatrix} j_1 & j_2 & j \\ m_{j1} & m_{j2} & -m_j \end{pmatrix}$ is not a matrix, but it is a Wigner $3j$-Symbol that can be calculated using math software.

As an example, consider a molecule with two valence electrons. Since the electrons are spin-$\frac{1}{2}$ particles, they can be paired ($s_1 = \frac{1}{2}$, $s_2 = -\frac{1}{2}$) so that the total spin, $S = s_1 + s_2 = 0$. The *multiplicity* of this state is defined as $(2S + 1) = 1$ and the state is called a *singlet state*. Many stable molecules have singlets as their ground electronic states. The first excited state of most stable molecules is a *triplet* state, rather than a singlet state. A state is a triplet if the two valence electrons are unpaired ($s_1 = \frac{1}{2}$, $s_2 = \frac{1}{2}$) so that the total spin, $S = s_1 + s_2 = 1$. The multiplicity becomes $(2S + 1) = 3$, hence the name, triplet state.

Let's write down the possible kets for the triplet and singlet states formed by two electrons (spin-$\frac{1}{2}$ particles) using our ket notation, where $|j, m_j\rangle = |S, M_S\rangle$. From now on, we will write capital letters to describe angular momenta from the coupling of two or more sources of angular momenta.

$$j_1 = \frac{1}{2} \quad j_2 = \frac{1}{2} \quad J = 1, 0 \quad (3.37)$$

$$s_1 = \frac{1}{2} \quad s_2 = \frac{1}{2} \quad S = 1, 0 \quad (3.38)$$

Now, we can use our Wigner $3j$-Symbols to translate the mathematical language between the coupled and uncoupled pictures:

$$|S, M_S\rangle = \sum_{m_{s1}, m_{s2}} W3j(s_1, s_2, S; m_{s1}, m_{s2}, M_S)\, |m_{s1}; m_{s2}\rangle \quad (3.39)$$

The Wigner $3j$-Symbols are written as follows, where s_1 and s_2 are coupled together to give S.

$$W3j(s_1, s_2, S; m_{s1}, m_{s2}, M_S) =$$

$$(-1)^{s_1 - s_2 + M_S} \sqrt{2S + 1} \begin{pmatrix} s_1 & s_2 & S \\ m_{s1} & m_{s2} & -M_S \end{pmatrix} \quad (3.40)$$

First, consider the triplet state. There is only one combination of m_{s1}, m_{s2} values that will give $M_S = +1$. In this case,

$$W3j(.5, .5, 1; .5, .5, 1) = (-1)^{.5 - .5 + 1} \sqrt{2(1) + 1} \begin{pmatrix} .5 & .5 & 1 \\ .5 & .5 & -1 \end{pmatrix} = 1 \quad (3.41)$$

and

$$|1, +1\rangle = |\frac{1}{2}; \frac{1}{2}\rangle \quad (3.42)$$

There are two combinations of m_{s1}, m_{s2} that give rise to $M_S = 0$, and both of the corresponding Wigner $3j$-Symbols are evaluated as $\frac{1}{\sqrt{2}}$:

$$|1, 0\rangle = \frac{1}{\sqrt{2}} |\frac{1}{2}; -\frac{1}{2}\rangle + \frac{1}{\sqrt{2}} |-\frac{1}{2}; \frac{1}{2}\rangle \quad (3.43)$$

There is only one combination of m_{s1}, m_{s2} values that will give $M_S = -1$.

$$|1, -1\rangle = |-\frac{1}{2}; -\frac{1}{2}\rangle \quad (3.44)$$

For the singlet state, we also have two possible combinations of m_{s1}, m_{s2} values and corresponding Wigner $3j$-Symbols:

$$|0, 0\rangle = \frac{1}{\sqrt{2}} |\frac{1}{2}; -\frac{1}{2}\rangle - \frac{1}{\sqrt{2}} |-\frac{1}{2}; \frac{1}{2}\rangle \quad (3.45)$$

This discussion of the coupled vs. uncoupled pictures and triplet vs. singlet states is key to understanding hyperfine structure, the Zeeman effect, and spin-orbit interaction in atoms and molecules. We will see that these quantum phenomena are the basis for everyday, modern technology such as compact fluorescent lighting, atomic clocks and global positioning systems (GPS).

3.6 Hyperfine Structure of Atoms

3.6.1 What is Hyperfine Interaction?

Hyperfine structure [10, 11, 89] is the splitting of spectral lines due to the interaction of the electron spin with the electric and magnetic fields of the nucleus in an atom or molecule. These nuclear fields could be magnetic dipole or electric quadrupole fields, for example. The field of radio astronomy, atomic clock technology and GPS devices would not exist without hyperfine interactions in atoms.

3.6.2 Hydrogen Hyperfine Hamiltonian (Coupled Basis)

For the hydrogen atom in its ground $^2S_{\frac{1}{2}}$ state, we can write the matrix representation of the Hamiltonian in the coupled basis, and diagonalize it to calculate the splitting between the hyperfine levels in MHz or cm.

The hyperfine interaction arises from the nuclear spin $(i = \frac{1}{2})$ and electron spin $(s = \frac{1}{2})$ coupling to give total angular momentum, $F = 1, 0$.

$$F = i + s, \quad i + s - 1, \ldots |i - s| \tag{3.46}$$

The electron of hydrogen in its ground electronic state has no orbital angular momentum $(l = 0)$, so the magnetic field, B_e, felt by the proton is [11]:

$$B_e = -\frac{16\pi}{3} \mu_B |\Psi_{1s}(0)|^2 \hat{s} \tag{3.47}$$

where $\mu_B = \frac{e\hbar}{2m_e} = 1.40$ MHz/G. Given the fact that the square of the hydrogen $1s$ wavefunction at the nucleus is,

$$|\Psi_{1s}(0)|^2 = \frac{1}{\pi a_0^3} \tag{3.48}$$

where $a_0 = \frac{\hbar^2}{me^2} = 5.292 \times 10^{-9}$ cm is the Bohr radius, we can write

$$B_e = -\frac{16}{3a_0^3} \mu_B \hat{s} \tag{3.49}$$

The Hamiltonian, $\hat{H}^{hyf}$, that represents the coupling of the magnetic moment of the proton to the electron's magnetic field is [11]:

$$\hat{H}^{hyf} = -\mu_p \cdot B_e = \frac{16}{3a_0^3} g_p \mu_N \mu_B \hat{i} \cdot \hat{s} \tag{3.50}$$

where $g_p = 5.58$ is the proton g-factor and $\mu_N = \frac{e\hbar}{2m_p} = 762$ Hz/G is the nuclear magneton. We can set up this Hamiltonian in matrix form using the coupled basis set. The Hamiltonian becomes:

$$\hat{H}^{hyf} = a\hat{i} \cdot \hat{s} \tag{3.51}$$

where,

$$a = \frac{16}{3a_0^3} g_p \mu_N \mu_B = 1420 \, \text{MHz} \tag{3.52}$$

Since,

$$\hat{F}^2 = (\hat{i} + \hat{s}) \cdot (\hat{i} + \hat{s}) = \hat{i}^2 + \hat{s}^2 + 2\hat{i} \cdot \hat{s} \tag{3.53}$$

and

$$a\hat{i} \cdot \hat{s} = \frac{a(\hat{F}^2 - \hat{i}^2 - \hat{s}^2)}{2} \tag{3.54}$$

we can rewrite the Hamiltonian as:

$$\hat{H}^{hyf} = a\hat{i} \cdot \hat{s} = \frac{a}{2}(\hat{F}^2 - \hat{i}^2 - \hat{s}^2) \tag{3.55}$$

The matrix elements in the coupled basis are:

$$\langle F, M_F | \hat{H}^{hyf} | F', M_F' \rangle = \frac{a}{2} \langle F, M_F | \hat{F}^2 - \hat{i}^2 - \hat{s}^2 | F', M_F' \rangle \tag{3.56}$$

The matrix elements can be solved to give:

$$\langle F, M_F | \hat{H}^{hyf} | F', M_F' \rangle = \frac{a}{2}[F(F+1) - i(i+1) - s(s+1)]\delta_{F,F'}\delta_{M_F,M_F'} \tag{3.57}$$

All that is left to do is calculate the energy splitting between the two ground state hyperfine levels, $F = 0$ and $F = 1$. Plugging in $i = \frac{1}{2}$ and $s = \frac{1}{2}$, we obtain a splitting of 1420 MHz = 21 cm. We can also build the matrix in the $|F, M_F\rangle$ basis to see that our choice of the coupled picture generates a diagonal matrix with eigenvalues $\frac{a}{4}$ and $-\frac{3a}{4}$. The energy difference between the two levels is simply $a = 1420$ MHz = 21 cm.

Table 3.1: Hydrogen Hyperfine Hamiltonian (Coupled Basis)

$\lvert F, M_F \rangle$	$\lvert 1,1 \rangle$	$\lvert 1,-1 \rangle$	$\lvert 1,0 \rangle$	$\lvert 0,0 \rangle$
$\langle 1,1 \rvert$	$\frac{a}{4}$	0	0	0
$\langle 1,-1 \rvert$	0	$\frac{a}{4}$	0	0
$\langle 1,0 \rvert$	0	0	$\frac{a}{4}$	0
$\langle 0,0 \rvert$	0	0	0	$-\frac{3a}{4}$

Hyperfine Splitting

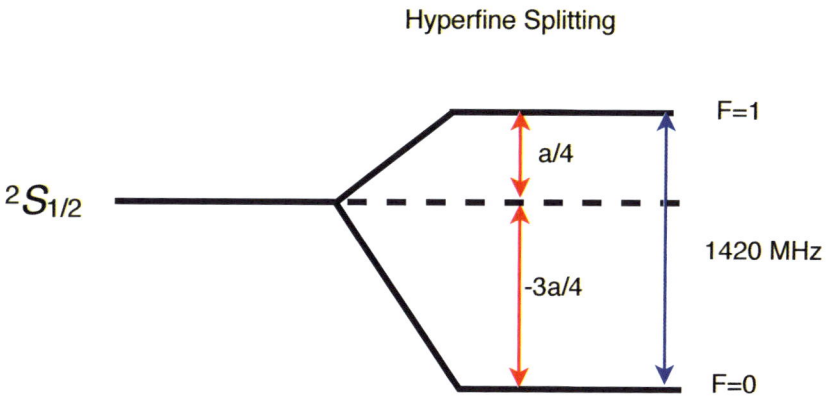

Figure 3.3: Hydrogen atom ground state hyperfine structure.

3.6.3 Hydrogen Hyperfine Hamiltonian (Uncoupled Basis)

We can now solve the same hydrogen hyperfine problem using the uncoupled picture. The hyperfine Hamiltonian in the uncoupled basis is:

$$\hat{H}^{hyf} = a\hat{i} \cdot \hat{s} = a(\hat{i}_z \cdot \hat{s}_z) + \frac{a(\hat{i}_+\hat{s}_- + \hat{i}_-\hat{s}_+)}{2} \tag{3.58}$$

where the first term, $a(\hat{i}_z \cdot \hat{s}_z)$ has only diagonal matrix elements, and the second term, $\frac{a(\hat{i}_+\hat{s}_- + \hat{i}_-\hat{s}_+)}{2}$, produces off-diagonal matrix elements between basis functions with the same value of $m_i + m_s$.

First, let's write out the uncoupled basis functions. In general, we have,

$$\psi_{s,m_s;i,m_i} = |s, m_s; i, m_i\rangle = |s, m_s\rangle |i, m_i\rangle = |m_s; m_i\rangle \tag{3.59}$$

where $s = \frac{1}{2}$ and $i = \frac{1}{2}$. There are four uncoupled functions, $|m_s; m_i\rangle$:

$$|\frac{1}{2}; \frac{1}{2}\rangle \quad |\frac{1}{2}; -\frac{1}{2}\rangle \quad |-\frac{1}{2}; \frac{1}{2}\rangle \quad |-\frac{1}{2}; -\frac{1}{2}\rangle \tag{3.60}$$

Next, we can explicitly write out the Hamiltonian matrix in the uncoupled basis set and evaluate the diagonal matrix elements and any non-zero off-diagonal matrix elements. The elements along the diagonal take the following form:

$$\langle m_s; m_i| \, a\hat{i}_z \cdot \hat{s}_z \, |m'_s; m'_i\rangle = am_i m_s \delta_{m_i, m'_i} \delta_{m_s, m'_s} \tag{3.61}$$

The off-diagonal matrix elements have the following form:

$$\langle m_s; m_i| \frac{a(\hat{i}_+\hat{s}_- + \hat{i}_-\hat{s}_+)}{2} |m_s; m_i\rangle = \frac{a}{2} \tag{3.62}$$

The full Hamiltonian matrix is:

Table 3.2: Hydrogen Hyperfine Hamiltonian (Uncoupled Basis)

$\langle m_s; m_i \vert \hat{H} \vert m_s; m_i \rangle$	$\vert \frac{1}{2}; \frac{1}{2} \rangle$	$\vert \frac{1}{2}; -\frac{1}{2} \rangle$	$\vert -\frac{1}{2}; \frac{1}{2} \rangle$	$\vert -\frac{1}{2}; -\frac{1}{2} \rangle$
$\langle \frac{1}{2}; \frac{1}{2} \vert$	$\frac{a}{4}$	0	0	0
$\langle \frac{1}{2}; -\frac{1}{2} \vert$	0	$-\frac{a}{4}$	$\frac{a}{2}$	0
$\langle -\frac{1}{2}; \frac{1}{2} \vert$	0	$\frac{a}{2}$	$-\frac{a}{4}$	0
$\langle -\frac{1}{2}; -\frac{1}{2} \vert$	0	0	0	$\frac{a}{4}$

When this matrix is diagonalized to obtain the eigenvalues, the result for $^2S_{\frac{1}{2}}$ $F = 1$ is,

$$E = \frac{a}{4} \qquad (3.63)$$

The result for $^2S_{\frac{1}{2}}$ $F = 0$ is,

$$E = \frac{-3a}{4} \qquad (3.64)$$

These are the same eigenvalues that we obtained using the coupled picture. The matrix representation in either basis will give us the correct answer. The only difference is that one choice of basis gives us a Hamiltonian that is already diagonal and the other does not. In quantum matrix mechanics, an important part of problem solving involves choosing a basis so that the Hamiltonian matrix representation is most nearly diagonal.

3.7 The Zeeman Effect

3.7.1 What is the Zeeman Effect?

The Zeeman effect is the interaction of an atom or molecule with a magnetic field [87,89,91]. The $\hat{H}^{Zee}$ operator for the interaction of the magnetic moment ($\hat{\mu}_M$) of hydrogen's electron with an external magnetic field (B) can be written as:

$$\hat{H}^{Zee} = -\hat{\mu}_M \cdot B = g_s \mu_B B \hat{s}_z \qquad (3.65)$$

where $g_s \approx 2$ is the g-value of the electron. The matrix representation of $\hat{H}^{Zee}$ is diagonal in the uncoupled basis ($|s, m_s; i, m_i\rangle$), but not in the coupled basis ($|i, s; F, M_F\rangle$), unlike the matrix representation for $\hat{H}^{hyf}$.

3.7.2 Hydrogen Zeeman Hamiltonian (Coupled Basis)

In order to set up the total Hamiltonian matrix for the hydrogen atom in a magnetic field, we must add the hyperfine and Zeeman Hamiltonian matrices, $\hat{H}^T = \hat{H}^{hyf} + \hat{H}^{Zee}$, and diagonalize to get the energies. We can then plot the energies as a function of magnetic field.

We already have the matrix representation of $\hat{H}^{hyf}$ in the coupled basis (table 3.1). However, the coupled basis functions, $|F, M_F\rangle$ are not eigenfunctions of the $\hat{H}^{Zee}$ that we wrote down in equation (3.65). Only the uncoupled basis functions, $|m_s; m_i\rangle$ are eigenfunctions of $\hat{H}^{Zee}$ in equation (3.65). We must use our Wigner 3j-Symbols to find a representation for $\hat{H}^{Zee}$ in the coupled picture. Remember that,

$$|F, M_F\rangle = \sum_{m_s, m_i} W3j(s, i, F; m_s, m_i, M_F) |m_s; m_i\rangle \qquad (3.66)$$

Now, we can write our basis functions as,

$$|1, +1\rangle = |\tfrac{1}{2}; \tfrac{1}{2}\rangle \qquad (3.67)$$

$$|1, 0\rangle = \frac{1}{\sqrt{2}} |\tfrac{1}{2}; -\tfrac{1}{2}\rangle + \frac{1}{\sqrt{2}} |-\tfrac{1}{2}; \tfrac{1}{2}\rangle \qquad (3.68)$$

$$|1, -1\rangle = |-\tfrac{1}{2}; -\tfrac{1}{2}\rangle \qquad (3.69)$$

$$|0, 0\rangle = \frac{1}{\sqrt{2}} |\tfrac{1}{2}; -\tfrac{1}{2}\rangle - \frac{1}{\sqrt{2}} |-\tfrac{1}{2}; \tfrac{1}{2}\rangle \qquad (3.70)$$

In order to calculate the matrix elements, let's first evaluate the possible eigenvalue equations. First, take $\hat{H}^{Zee} = 2\mu_B B \hat{s}_z$ and operate on $|1,1\rangle$ in uncoupled form:

$$2\mu_B B \hat{s}_z |1,1\rangle = 2\mu_B B \hat{s}_z |\frac{1}{2};\frac{1}{2}\rangle = 2\mu_B B \frac{1}{2}|\frac{1}{2};\frac{1}{2}\rangle = \mu_B B |\frac{1}{2};\frac{1}{2}\rangle \tag{3.71}$$

Next, take $\hat{H}^{Zee} = 2\mu_B B \hat{s}_z$ and operate on $|1,-1\rangle$ in uncoupled form:

$$2\mu_B B \hat{s}_z |1,-1\rangle = \tag{3.72}$$

$$2\mu_B B \hat{s}_z |-\frac{1}{2};-\frac{1}{2}\rangle = 2\mu_B B \left(-\frac{1}{2}\right)|-\frac{1}{2};-\frac{1}{2}\rangle = -\mu_B B |-\frac{1}{2};-\frac{1}{2}\rangle$$

Now, take $\hat{H}^{Zee} = 2\mu_B B \hat{s}_z$ and operate on $|1,0\rangle$ in uncoupled form:

$$2\mu_B B \hat{s}_z |1,0\rangle = 2\mu_B B \hat{s}_z \left[\frac{1}{\sqrt{2}}|\frac{1}{2};-\frac{1}{2}\rangle + \frac{1}{\sqrt{2}}|-\frac{1}{2};\frac{1}{2}\rangle\right] \tag{3.73}$$

$$= \frac{1}{\sqrt{2}}2\mu_B B \hat{s}_z \left[|\frac{1}{2};-\frac{1}{2}\rangle + |-\frac{1}{2};\frac{1}{2}\rangle\right]$$

$$= \frac{1}{\sqrt{2}}2\mu_B B \left[\frac{1}{2}|\frac{1}{2};-\frac{1}{2}\rangle + -\frac{1}{2}|-\frac{1}{2};\frac{1}{2}\rangle\right]$$

$$= \frac{1}{\sqrt{2}}\frac{2}{2}\mu_B B \left[|\frac{1}{2};-\frac{1}{2}\rangle - |-\frac{1}{2};\frac{1}{2}\rangle\right]$$

$$= \mu_B B \left[\frac{1}{\sqrt{2}}|\frac{1}{2};-\frac{1}{2}\rangle - \frac{1}{\sqrt{2}}|-\frac{1}{2};\frac{1}{2}\rangle\right]$$

$$= \mu_B B |0,0\rangle$$

This results in a non-zero, off-diagonal matrix element.

Finally, take $\hat{H}^{Zee} = 2\mu_B B \hat{s}_z$ and operate on $|0,0\rangle$ in uncoupled form:

$$2\mu_B B \hat{s}_z |0,0\rangle = 2\mu_B B \hat{s}_z \left[\frac{1}{\sqrt{2}}|\frac{1}{2};-\frac{1}{2}\rangle - \frac{1}{\sqrt{2}}|-\frac{1}{2};\frac{1}{2}\rangle\right] \tag{3.74}$$

$$= \frac{1}{\sqrt{2}} 2\mu_B B \hat{s}_z \left[|\frac{1}{2}; -\frac{1}{2}\rangle - |-\frac{1}{2}; \frac{1}{2}\rangle \right]$$

$$= \frac{1}{\sqrt{2}} 2\mu_B B \left[\frac{1}{2} |\frac{1}{2}; -\frac{1}{2}\rangle - -\frac{1}{2} |-\frac{1}{2}; \frac{1}{2}\rangle \right]$$

$$= \frac{1}{\sqrt{2}} \frac{2}{2} \mu_B B \left[|\frac{1}{2}; -\frac{1}{2}\rangle + |-\frac{1}{2}; \frac{1}{2}\rangle \right]$$

$$= \mu_B B \left[\frac{1}{\sqrt{2}} |\frac{1}{2}; -\frac{1}{2}\rangle + \frac{1}{\sqrt{2}} |-\frac{1}{2}; \frac{1}{2}\rangle \right]$$

$$= \mu_B B |1,0\rangle$$

Again, we have a non-zero, off-diagonal matrix element.
    Now, we can build the total Hamiltonian:

Table 3.3: Ground State Hydrogen Hyperfine-Zeeman Hamiltonian (Coupled Basis)

$|F, M_F\rangle$	$|1,1\rangle$	$|1,-1\rangle$	$|1,0\rangle$	$|0,0\rangle$
$\langle 1,1|$	$\frac{a}{4} + \mu_B B$	0	0	0
$\langle 1,-1|$	0	$\frac{a}{4} - \mu_B B$	0	0
$\langle 1,0|$	0	0	$\frac{a}{4}$	$\mu_B B$
$\langle 0,0|$	0	0	$\mu_B B$	$-\frac{3a}{4}$

    Finally, we can diagonalize the matrix to obtain eigenvalues, and plot the energies as a function of magnetic field. Here is the result:

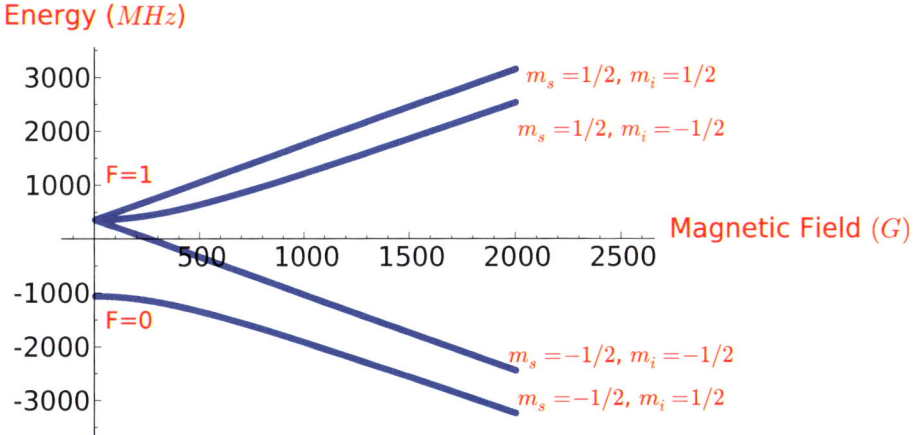

Figure 3.4: Hydrogen atom Zeeman effect.

Notice that at low magnetic field, s, i, F, M_F are *good* quantum numbers, since s and i are coupled together due to the hyperfine interaction to give $F = 0, 1$. For weak magnetic fields near 0 G, the levels tune linearly with field.

At intermediate fields, near 500 G, you can see that the off-diagonal matrix elements mix the two $M_F = 0$ levels and push them apart from one another, and the levels tune quadratically with field. The levels and quantum numbers are all mixed up at these field strengths.

As the applied magnetic field gets stronger than 500 G, the splitting between the individual sublevels gets larger than the 1420 MHz hyperfine splitting. As this happens, the coupling of s and i to the magnetic field becomes stronger than the hyperfine interaction, and the dependence on magnetic field is linear once again. The good quantum numbers at these high fields are the uncoupled s, m_s, i, m_i.

3.7.3 Hydrogen Zeeman Hamiltonian (Uncoupled Basis)

Solving the hydrogen atom hyperfine-Zeeman Hamiltonian in the uncoupled basis is straightforward. Again, we will add the hyperfine and Zeeman Hamiltonian matrices, $\hat{H}^{hyf} + \hat{H}^{Zee}$, to obtain the total matrix. We already know the hyperfine matrix in the uncoupled basis. The Zeeman matrix elements are easy to calculate, because the uncoupled basis is an eigenfunction of our Zeeman operator,

$$\hat{H}^{Zee} = g_s \mu_B B \hat{s}_z \tag{3.75}$$

and the Zeeman matrix is diagonal. The total matrix is:

Table 3.4: Ground State Hydrogen Hyperfine-Zeeman Hamiltonian (Uncoupled Basis)

$|m_s; m_i\rangle$	$|\frac{1}{2}; \frac{1}{2}\rangle$	$|\frac{1}{2}; -\frac{1}{2}\rangle$	$|-\frac{1}{2}; \frac{1}{2}\rangle$	$|-\frac{1}{2}; -\frac{1}{2}\rangle$
$\langle\frac{1}{2}; \frac{1}{2}|$	$\mu_B B_z + \frac{a}{4}$	0	0	0
$\langle\frac{1}{2}; -\frac{1}{2}|$	0	$\mu_B B_z - \frac{a}{4}$	$\frac{a}{2}$	0
$\langle-\frac{1}{2}; \frac{1}{2}|$	0	$\frac{a}{2}$	$-\mu_B B_z - \frac{a}{4}$	0
$\langle-\frac{1}{2}; -\frac{1}{2}|$	0	0	0	$-\mu_B B_z + \frac{a}{4}$

Try diagonalizing this matrix and plot the energies as a function of field.

3.8   The Stark Effect

The Stark effect [87, 89] describes the interaction of an atom or molecule with an electric field, E:

$$\hat{H}^{Stk} = -\hat{\mu}_E \cdot E \qquad (3.76)$$

The same quantum matrix mechanics methods that we used in treating the Zeeman operator can be applied to the Stark operator. As an atomic example, figure 3.5 shows a calculation of the hyperfine structure and Stark splitting in the spectrum of rubidium-87 for the D_2 $4p^6 5p \left({}^2P_{\frac{3}{2}} \right) \leftarrow 4p^6 5s \left({}^2S_{\frac{1}{2}} \right)$ transition near 780 nm.

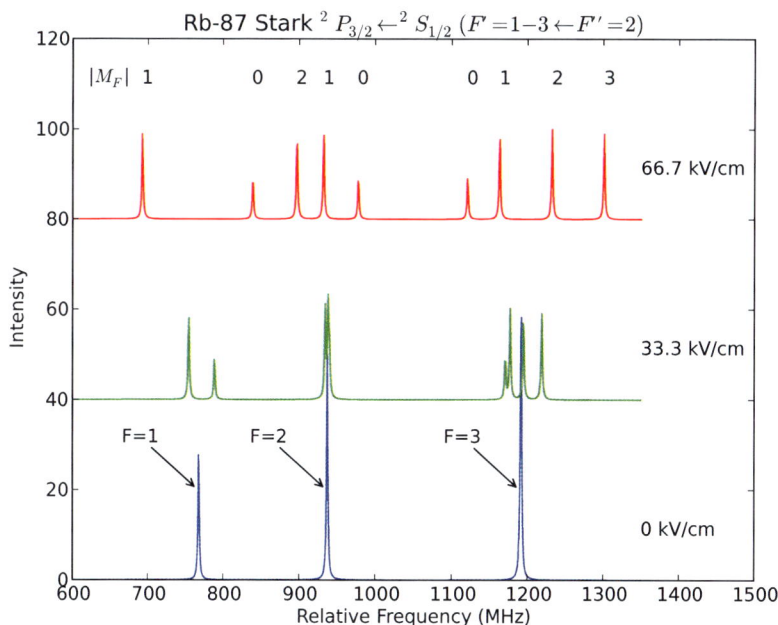

Figure 3.5: Simulated Stark effect spectra of rubidium-87 in different electric fields.

The Stark effect is most useful as an experimental probe of structure and bonding in molecules. Take a look at some reference articles [72–79, 90, 92] on the molecular Stark effect to learn about the power of Stark spectroscopy in physical chemistry.

3.9 Spin-Orbit Effects

3.9.1 Energy Levels of the Hydrogen Atom

So far, we have only considered the ground electronic state of the hydrogen atom with one electron in a $1s$ orbital. The term symbol for this state is written as $^2S_{\frac{1}{2}}$. In general, term symbols are written as $^{(2S+1)}L_J$. $(2S+1)$ is the multiplicity, L is the orbital angular momentum, and $J = L+S$ is the total angular momentum, calculated using the Clebsch-Gordan series. The first excited state of the hydrogen atom is obtained by exciting the electron in the $1s$ orbital to a $2p$ orbital. The $2p$ electronic configuration gives rise to a ^2P term symbol with two possible values of J, corresponding to two different energies:

$$E(^2P_{\frac{1}{2}}) = 82,258.92\,\text{cm}^{-1} \tag{3.77}$$

$$E(^2P_{\frac{3}{2}}) = 82,259.28\,\text{cm}^{-1} \tag{3.78}$$

The ^2P state is split (by .36 cm^{-1}) into two levels by the spin-orbit interaction.

3.9.2 Spin-Orbit Hamiltonian

As you know, electrons are charged particles. The orbital motion of the charged electron in the hydrogen atom produces a current, and gives rise to a magnetic field. Electrons also have spin that gives rise to a magnetic moment. Spin-orbit coupling [3] is the interaction between the magnetic moments produced by the orbital motion and spin of the electron.

The magnitude of the electron spin magnetic moment is proportional to $\hat{s}$,

$$\hat{\mu}_s = -g_e \mu_B \hat{s} \tag{3.79}$$

and the magnitude of the electron orbital magnetic moment is proportional to $\hat{l}$,

$$\hat{\mu}_l = -g_l \mu_B \hat{l} \tag{3.80}$$

The spin-orbit Hamiltonian is:

$$\hat{H}^{s.o.} = \xi(r)\hat{l} \cdot \hat{s} \tag{3.81}$$

where,

$$\xi(r) = \frac{1}{2\mu^2 c^2} \frac{1}{r} \frac{\partial V}{\partial r} \qquad (3.82)$$

$$V = \frac{-Ze^2}{4\pi\epsilon_0 r} \qquad (3.83)$$

and V is the Coulomb potential between the electron and nucleus. We can now solve the spin-orbit Schrödinger equation for hydrogen in the ^2P state:

$$\hat{H}^{s.o.} \psi = E\psi \qquad (3.84)$$

3.9.3   The Coupled Representation

We can use the Clebsch-Gordan series to write down the quantum numbers associated with a $2p$ electron, where $l = 1$ and $s = \frac{1}{2}$.

$$J = l + s \qquad J = \frac{3}{2}, \frac{1}{2} \qquad (3.85)$$

The coupled basis is written as:

$$\psi_{l,s;J,M_J} = |l, s; J, M_J\rangle = |J, M_J\rangle \qquad (3.86)$$

There are six basis functions:

$$\left|\frac{3}{2}, \frac{3}{2}\right\rangle \quad \left|\frac{3}{2}, \frac{1}{2}\right\rangle \quad \left|\frac{3}{2}, -\frac{1}{2}\right\rangle \quad \left|\frac{3}{2}, -\frac{3}{2}\right\rangle \qquad (3.87)$$

$$\left|\frac{1}{2}, \frac{1}{2}\right\rangle \qquad \left|\frac{1}{2}, -\frac{1}{2}\right\rangle$$

If we represent the $\hat{H}^{s.o.}$ operator in terms of $\hat{J}^2$, $\hat{l}^2$ and $\hat{s}^2$ in the coupled basis, the Hamiltonian matrix will already be diagonal, because the coupled basis is an eigenfunction of these squared operators.

$$\hat{J}^2 = (\hat{l} + \hat{s}) \cdot (\hat{l} + \hat{s}) = \hat{l}^2 + \hat{s}^2 + 2\hat{l} \cdot \hat{s} \qquad (3.88)$$

since $\hat{l}$ and $\hat{s}$ commute. We can rewrite the spin-orbit operator as:

$$\xi(r)\hat{l} \cdot \hat{s} = \frac{\xi(r)(\hat{J}^2 - \hat{l}^2 - \hat{s}^2)}{2} \qquad (3.89)$$

and calculate matrix elements:

$$E = \langle J, M_J | \xi(r) \hat{l} \cdot \hat{s} | J', M_J' \rangle \tag{3.90}$$

$$E = \frac{1}{2} \langle J, M_J | \xi(r) (\hat{J}^2 - \hat{l}^2 - \hat{s}^2) | J', M_J' \rangle \tag{3.91}$$

$$E = \frac{1}{2} \langle J, M_J | \xi(r) | J', M_J' \rangle \langle J, M_J | (\hat{J}^2 - \hat{l}^2 - \hat{s}^2) | J', M_J' \rangle \tag{3.92}$$

The matrix element on the left is an integral over the radial part of the hydrogen atomic wavefunction:

$$\langle J, M_J | \xi(r) | J, M_J \rangle = \int R_{2p}^*(r) \xi(r) R_{2p}(r) r^2 dr = \zeta_{2p} = 0.24 \, \text{cm}^{-1} \tag{3.93}$$

The matrix element on the right is evaluated as:

$$\langle J, M_J | (\hat{J}^2 - \hat{l}^2 - \hat{s}^2) | J', M_J' \rangle \tag{3.94}$$

$$= [J(J+1) - l(l+1) - s(s+1)] \delta_{J,J'} \delta_{M_J, M_J'}$$

The spin-orbit energy becomes:

$$E = \frac{\zeta_{2p}}{2} [J(J+1) - l(l+1) - s(s+1)] \tag{3.95}$$

Plugging in the values for the quantum numbers,

$$l = 1, s = \frac{1}{2}, J = \frac{3}{2}, \frac{1}{2} \tag{3.96}$$

we obtain two energy levels,

$$E(^2P_{\frac{3}{2}}) = \frac{\zeta_{2p}}{2} \tag{3.97}$$

$$E(^2P_{\frac{1}{2}}) = -\zeta_{2p} \tag{3.98}$$

with a spin-orbit splitting of $\frac{3\zeta_{2p}}{2} = .36 \, \text{cm}^{-1}$ for hydrogen in its ^2P state.

3.9.4 The Uncoupled Representation

Now, we can solve the spin-orbit problem for an electron in the $2p$ ^2P state of hydrogen using the uncoupled representation. The spin-orbit Hamiltonian in the uncoupled basis is,

$$\hat{H}^{s.o.} = \xi(r)\hat{l} \cdot \hat{s} = \xi(r)\hat{l}_z \cdot \hat{s}_z + \frac{\xi(r)(\hat{l}_+\hat{s}_- + \hat{l}_-\hat{s}_+)}{2} \tag{3.99}$$

where the first term, $\xi(r)\hat{l}_z \cdot \hat{s}_z$, has only diagonal matrix elements, and the second term, $\frac{\xi(r)(\hat{l}_+\hat{s}_- + \hat{l}_-\hat{s}_+)}{2}$, gives rise to off-diagonal matrix elements between basis functions with the same value of $m_l + m_s$.

The uncoupled basis functions for the $2p$ hydrogen atom take the form:

$$\psi_{l,m_l;s,m_s} = |l, m_l; s, m_s\rangle = |l, m_l\rangle |s, m_s\rangle = |m_l; m_s\rangle \tag{3.100}$$

There are six uncoupled $|m_l; m_s\rangle$ functions:

$$|1; \pm\frac{1}{2}\rangle \qquad |0; \pm\frac{1}{2}\rangle \qquad |-1; \pm\frac{1}{2}\rangle \tag{3.101}$$

Diagonal matrix elements have the form:

$$\langle m_l; m_s | \xi(r)\hat{l}_z \cdot \hat{s}_z | m_l; m_s\rangle = \zeta_{2p} m_l m_s \tag{3.102}$$

The non-zero off-diagonal matrix elements have the form:

$$\frac{\zeta_{2p}}{2} \langle m_l; m_s | \hat{l}_+\hat{s}_- + \hat{l}_-\hat{s}_+ | m_l'; m_s'\rangle = \frac{1}{\sqrt{2}}\zeta_{2p} \tag{3.103}$$

As an example, let's calculate one of the off-diagonal matrix elements,

$$\frac{\zeta_{2p}}{2} \langle 0, \frac{1}{2} | \hat{l}_+\hat{s}_- + \hat{l}_-\hat{s}_+ | 1, -\frac{1}{2}\rangle \tag{3.104}$$

The first term is evaluated as,

$$\frac{\zeta_{2p}}{2} \langle 0, \frac{1}{2} | \hat{l}_+\hat{s}_- | 1, -\frac{1}{2}\rangle = 0 \tag{3.105}$$

because you can't lower the value of m_s any further. The second term can be evaluated as,

$$\frac{\zeta_{2p}}{2} \langle 0, \frac{1}{2} | \hat{l}_-\hat{s}_+ | 1, -\frac{1}{2}\rangle = \tag{3.106}$$

$$\frac{\zeta_{2p}}{2} \langle 0|\hat{l}_-|1\rangle \langle \frac{1}{2}|\hat{s}_+| - \frac{1}{2}\rangle = \tag{3.107}$$

$$\frac{\zeta_{2p}}{2} \sqrt{l(l+1) - m_l(m_l - 1)} \langle 0|0\rangle \sqrt{s(s+1) - m_s(m_s + 1)} \langle \frac{1}{2}\frac{1}{2}\rangle = \tag{3.108}$$

$$\frac{\zeta_{2p}}{2} \sqrt{2-0} \sqrt{\frac{3}{4} + \frac{1}{4}} = \frac{\zeta_{2p}\sqrt{2}}{2}$$

Now, we can explicitly write out the spin-orbit Hamiltonian matrix in the uncoupled basis set:

Table 3.5: Spin-Orbit Hamiltonian Matrix for Hydrogen $2p$

$\langle|\hat{H}|\rangle$	$|1,\frac{1}{2}\rangle$	$|1,-\frac{1}{2}\rangle$	$|0,\frac{1}{2}\rangle$	$|0,-\frac{1}{2}\rangle$	$|-1,\frac{1}{2}\rangle$	$|-1,-\frac{1}{2}\rangle$
$\langle 1,\frac{1}{2}|$	$\zeta_{2p}/2$	0	0	0	0	0
$\langle 1,-\frac{1}{2}|$	0	$-\zeta_{2p}/2$	$\zeta_{2p}\sqrt{2}/2$	0	0	0
$\langle 0,\frac{1}{2}|$	0	$\zeta_{2p}\sqrt{2}/2$	0	0	0	0
$\langle 0,-\frac{1}{2}|$	0	0	0	0	$\zeta_{2p}\sqrt{2}/2$	0
$\langle -1,\frac{1}{2}|$	0	0	0	$\zeta_{2p}\sqrt{2}/2$	$-\zeta_{2p}/2$	0
$\langle -1,-\frac{1}{2}|$	0	0	0	0	0	$\zeta_{2p}/2$

This Hamiltonian matrix can be diagonalized to give the same eigenvalues that we obtained using the coupled picture:

$$E(^2P_{\frac{3}{2}}) = \frac{\zeta_{2p}}{2} \tag{3.109}$$

$$E(^2P_{\frac{1}{2}}) = -\zeta_{2p} \tag{3.110}$$

Chapter 4

Hydrogen: Astronomy to Aliens

4.1 The Birth of Radio Astronomy

The mathematics and angular momentum theory of hydrogen hyperfine structure are not just some esoteric quantum mechanical subjects. It may come as a surprise to you that much of what we know about our universe rests on the ability in astronomy to understand and measure the 1420 MHz hyperfine splitting in the ground electronic state of the hydrogen atom. Observation of this splitting marked the birth of the entire field of radioastronomy [64]. Much of what we know about the Big Bang, black holes and the structure of galaxies has come from radioastronomy measurements, rather than optical measurements.

In 1931, Karl G. Jansky of Bell Labs built a rotating directional radio antenna in order to discover the sources of long wavelength radio static that might interfere with ship-to-shore trans-Atlantic communications. Jansky identified some radio static that was due to thunderstorms. More importantly, some of the static was due to extraterrestrial radio waves from the center of our Milky Way galaxy. Stars were emitting radio waves! Jansky's discovery was published in the New York Times in 1933 [1]. This was the beginning of radioastronomy, where matter in the universe is sending out radio signals for us to listen to with our antennas.

Most of the matter in the universe is made of hydrogen gas. Stars and the interstellar medium between stars are mostly hydro-

71

gen atoms, ions, or molecules. If we want to learn about the matter in our universe, we should try to detect the hydrogen. Toward the end of World War II, an astronomer named Hendrik C. van de Hulst suggested a way to detect hydrogen atoms in space using the 1420 MHz = 21 cm hyperfine transition. Any hydrogen atoms that might be excited from the $F = 0 \rightarrow F = 1$ hyperfine level would emit a radio signal at 21 cm as they dropped back down to $F = 0$.

Harvard physicists Edward M. Purcell (who shared the Nobel Prize for measuring nuclear magnetic resonance) and Harold I. Ewen were the pioneers in building an experiment to find the signature of hydrogen atoms in outer space [26]. With a budget of only a few hundred dollars, Purcell and Ewen built an antenna in 1951, and detected the 21 cm radio signal from hydrogen.

4.2 Mapping the Universe with Hydrogen

Knowledge of the hyperfine transition in hydrogen and detection via radio antennas have led to amazing discoveries about galaxies. Using the Doppler effect, astronomers have proved that our own Milky Way galaxy is spiral, and the galactic rotation has been analyzed due to the fact that the hydrogen atoms moving away from the radio antenna result in a longer wavelength signal, while the hydrogen atoms moving toward the antenna result in a signal shift to shorter wavelengths. The NRAO (National Radio Astronomy Observatory) [57], a facility of the NSF (National Science Foundation) has a web page with images, podcasts and videos on how radio astronomy has revealed this hidden structure of our universe. In addition, NASA's Goddard Space Flight Center web page on the multi-wavelength Milky Way [56] has spectacular images of the Milky Way at different wavelengths.

4.3 Is Anyone(thing) Out There?

Many molecules have now been discovered in outer space through the use of spectroscopy in astronomy, including H_2O and NH_3. Organic molecules such as formaldehyde (H_2CO) and glycine (the simplest amino acid, NH_2CH_2COOH) have been found in space [24]. In this way, our understanding of quantum mechanics is key in guiding the search for extraterrestrial life. Finding molecules such as water on exoplanets can give us clues to possible existence of life on other worlds.

Chapter 5

Time and Position with Atomic Clocks

5.1 Clocks

5.1.1 What is a Clock?

A clock is simply an oscillator that runs at a particular frequency in order to measure time. Any good clock must have a frequency that is constant anywhere in the world at any time. Accuracy and precision in measuring time are extremely important in everyday life. Time synchronization is key for everything from email and text messages to TV broadcast signals, electronic banking transactions, radar and sonar for navigation, the Global Positioning System (GPS) and measuring magnetic and gravitational fields for medical and security applications and autonomous navigation. Many physical quantities, like the meter and the volt rely on the definition of the second as a standard unit of time.

We know that time is not really absolute, because space and time are inextricably linked. Einstein's theory of special relativity says that time depends on relative motion, while the theory of general relativity describes the effect of gravity on time. However, we don't need to worry about the space-time continuum for most everyday time measurements on earth.

The earliest kinds of clocks made use of the sun, stars and planets for short and long term timekeeping and navigation. You are probably familiar with sundials, hourglasses and pendulum-based clocks. The ancient Egyptians even made water clocks based on regulating the flow of water from one vessel to another.

Most modern electronic clocks are based on the quartz oscillator. Quartz-based clocks measure time through the regular vibrations of a quartz crystal that have been excited by an electric current. This phenomenon of a crystal vibrating when an AC voltage is applied is called the piezoelectric effect. One main limitation of the accuracy and precision of quartz oscillators has to do with the fact that the frequency is temperature dependent, and may be affected by impurities in the quartz crystal. In order to build a more precise clock that is independent of external effects such as temperature and internal effects such as composition, improved technology is necessary.

5.1.2 Atomic Clocks

How would you make a clock so good that it would not gain or lose a second in about 400 million years? You would have to build an atomic clock [89]. Atomic clocks are based on the transition frequency between hyperfine levels in atoms. In the same way that the pendulum in a grandfather clock swings back and forth at a particular frequency, in an atomic hyperfine transition, the electron spin exchanges angular momentum with the nuclear spin a fixed number of times a second.

The hyperfine structure of the Cs atom is very important for timekeeping, because the definition of the second, and the official United States clock, are based on the transition between the $F = 3$ and $F = 4$ hyperfine levels in the $6^2S_{\frac{1}{2}}$ electronic ground state of Cs. The Cs hyperfine structure is due to the ^{133}Cs nucleus with $I = \frac{7}{2}$ interacting with the electron spin. The $F = 3 \leftrightarrow F = 4$ transition has a frequency of 9.192631770 GHz. In order to increase the precision of the clock, a magnetic field can be applied to lift the degenerate F levels into M_F components. The M_F sublevels that exhibit the smallest Zeeman effect can be state-selected and used to measure time to high precision, without the influence of external conditions such as temperature. Since the atomic clock is based on a fundamental transition frequency between the quantized energy levels of an atom, this provides an excellent standard for measuring time. All Cs atoms in the universe are exactly the same, and have the potential to report their 9.192631770 GHz frequency for use as an atomic clock.

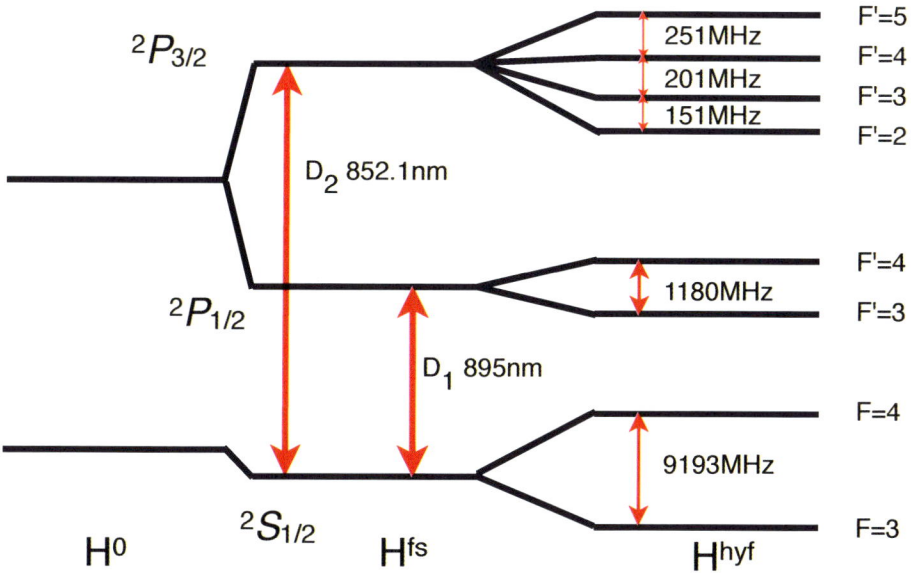

Figure 5.1: ^{133}Cs $(6^2P - 6^2S)$ atomic transitions with fine and hyperfine structure splitting.

NIST (National Institute of Standards and Technology) uses their NIST-F1 Cs fountain atomic clock [49] as the standard U.S. clock that defines the second as a unit of time. The clock has a fountain of Cs atoms inside a low-pressure vacuum chamber. Six infrared laser beams are oriented and tuned in such a way to cool and slow the Cs atoms very close to absolute zero into a tiny spherical cloud. The ultra-cold Cs atoms are directed into a microwave cavity using additional lasers where they are excited by microwaves. The atoms are then excited by yet another laser to make them fluoresce, and the microwave frequency, 9.192631770×10^9 Hz, is measured.

Many other types of atomic clocks have been made [47]. The hydrogen atomic maser clock is based on hydrogen hyperfine structure, and the rubidium clock is based on rubidium hyperfine structure [49]. The strontium clock [48] traps a few thousand strontium atoms in an optical lattice formed by the use of standing waves of near-IR laser light. An ultra-stable red laser drives the clock transition between strontium energy levels.

A mercury clock [58] has been built, based on a single mercury ion that will not gain or lose a second in 400 million years. The single ion clock provides higher precision than other atomic clocks that use tiny clouds of atoms, because the different parts of those atomic clouds may be affected by variations in electric and magnetic fields. The single ion in the mercury clock can be isolated in space using an electromagnetic trap and cooled with lasers for precise control. Light from a UV laser is used to excite a transition between two quantized energy levels, and this frequency is used for the oscillations of the clock. The oscillations of UV light are much faster than the oscillations of microwave light used in the Cs clock, so this leads to improved precision of the mercury clock.

The quantum logic clock [17] is based on a single aluminum atom, and became the world's most precise clock in 2010. It would neither lose or gain a second in about 3.7 billion years. The aluminum ion is trapped by electric fields and excited by UV light. The clock makes use of quantum computing to keep time.

A portable, potentially battery-operated atomic clock called a chip-scale clock [42, 69] that is about the size of a grain of rice, has been constructed. The clock is based on cesium, but is designed differently from the fountain clock and is also less precise than the fountain clock. In the chip-scale clock, cesium is confined in a cell and excited with an infrared laser. This smaller, cheaper atomic clock could become a more precise alternative to quartz oscillator clocks.

5.2 Global Positioning System (GPS)

You probably use or own a GPS personal navigation device that is integrated into your car or smart phone. The GPS receiver in your phone can calculate its position with an uncertainty less than about 10 meters. This is truly amazing! How is your phone able to calculate its exact position anywhere in the world? Atomic clocks hold the key to answering this question.

In order to determine the relative position of your GPS-enabled phone on a map, it is necessary to calculate a vector (direction and range). We must discuss how direction finding and ranging is performed [49]. Radio signals are good for more than communication alone. A simple loop antenna can be used to get one's bearings with respect to a broadcast station, or radio signal source, as long as the source is within the spatial range and frequency tuning range of the receiver. Take a look at the loop antenna in figure 5.2.

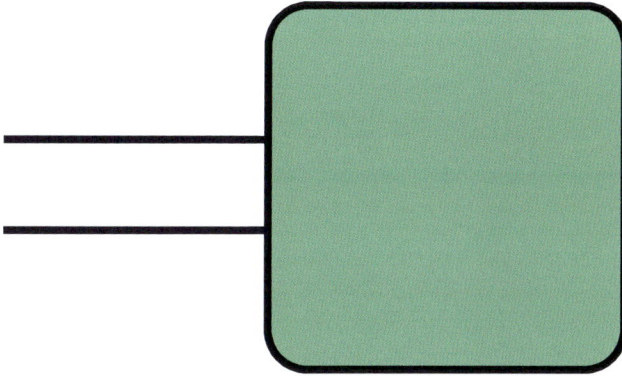

Figure 5.2: A loop antenna.

If the plane of the loop antenna is exactly perpendicular to the direction of the electric field of the radio wave, then no signal will be detected. However, if the plane of the loop is parallel to the direction of the electric field of the radio wave, the signal will be at a maximum. Rotation of the loop to tune through the maximum and minimum signal can be used to get one's bearings relative to the transmitting station.

Radar (RAdio Detection And Ranging) was invented at MIT just before WWII as a way of detecting enemy planes. Radar is a way of detecting the echo from a radio signal bounced off of a target, in the same way that bats use sonar to fly in the dark. If you can precisely measure the delay between the transmission and the reception of a radar signal that has reflected off of a target, you can calculate the distance to that target because the speed of light is a constant. You often hear of Doppler radar used by police in radar guns, or on TV as weather radar. Doppler radar allows one to calculate the direction, range, and velocity of a moving target. You know that a moving object would reflect a Doppler shifted frequency. The frequency shift in the reflected signal can be converted to a velocity for the moving target.

The Global Positioning System (GPS) is really made up of a constellation of satellites run by the U.S. Department of Defense (DOD). The GPS network is both a radio-navigation system and provides a standard for time and frequency. Each GPS satellite has either a cesium or rubidium atomic clock on board. The reception antenna in your smart phone must have line-of-sight to the GPS satellites. In other words, the antenna in your phone must have a clear path through the sky to receive the radio signals from the GPS satellites. As receivers, our GPS-enabled smart phones contain an antenna and a quartz clock. The phone can simultaneously receive signals from several different GPS satellites that are in sync with each other, and solve the equations to obtain coordinates and time. The Cs and Rb atomic clocks on the GPS satellites are key, because very high precision in measuring the time delay between signals is necessary to achieve high precision (low uncertainty) in the position of the smart phone receiver. None of this would be possible without quantum mechanics and hyperfine structure. Before quantum mechanics, who would have thought that the tiny magnetic interactions between nuclei and electrons would give rise to such important, and useful, technology?

Chapter 6

Quantum Computers and Qubits

6.1   Classical vs. Quantum Computing

You should know that traditional, or classical, computers are based on the binary system of 0's and 1's. This two-state system of bits (binary digits) already appears similar to quantized states of atoms and molecules. Moore's Law states that processor complexity doubles about every 2 years. As complexity increases, computer components get smaller. Soon, computer components will reach the molecular size level, where quantum effects are unavoidable. How can we use quantum effects to our advantage in computing?

6.2   Qubit: A Quantum Bit

A *qubit* is made of a two-state space, or system, where a quantum computer can be based on two different inputs, states or energy levels. For example, let $|\psi_0\rangle$ and $|\psi_1\rangle$ be the orthonormal basis for the space. We can then define the qubit as $|\Psi\rangle = a|\psi_0\rangle + b|\psi_1\rangle$. In quantum computing, you would make $|\psi_0\rangle = |0\rangle$ and $|\psi_1\rangle = |1\rangle$. This could be spin up vs. spin down for a quantum particle, analogous to 0 and 1 on a classical computer. Ψ must be normalized so that $\langle\Psi|\Psi\rangle = 1$, or $|a|^2 + |b|^2 = 1$. The true eigenvalue of the wavefunction, $|\Psi\rangle$ is uncertain until the eigenvalue, 0 or 1, is measured.

6.3   Quantum Computing in Action

Quantum computers have the potential to compute in ways that a classical computer can not. Quantum superposition, uncertainty and entanglement may enhance cryptography applications, or lead to smaller, more efficient computers. Quantum computers may be able to solve problems that are costly, time consuming or impossible using a classical computer.

The quantum logic clock [17, 66] uses quantum computer logic. This clock uses two different kinds of ions (aluminum/beryllium or aluminum/magnesium) that are closely confined in an electro-magnetic trap and laser cooled to temperatures near absolute zero. Quantum computing techniques are used to communicate information between the aluminum ion and the beryllium or magnesium ion in a tandem detection approach. Information about the aluminum ion states is transferred to the beryllium or magnesium ion. The aluminum ion is the stable oscillator source, and it has two energy levels that are analogous to the 1 and 0 of a computer. However, the aluminum ion oscillations are not easily detected using laser techniques. Instead, the beryllium or magnesium ion reports whether the aluminum ion remained in the ground 0 state, or jumped to the 1 state.

Chapter 7

The Quantum Light of Your Life

7.1 How Does a Light Bulb Work?

Incandescent lights have a tungsten filament inside a glass bulb that is filled with inert gas, such as argon. White light is produced by resistive heating in applying a current to the filament through electrical connections at the base of the bulb. The hot, glowing filament gives of a broad spectrum of visible light that appears white to the eye. The incandescent bulb emission spectrum follows Planck's quantum blackbody radiation law that we learned about in chapter 1. We wouldn't fully understand how a light bulb turns on without quantum mechanics!

7.2 Lighting the Roadway with Sodium

Sodium vapor street lamps are used near astronomy observatories. These lamps use an electrical discharge in vaporized sodium metal to excite the atoms from their ground electronic state, $2p^6 3s \left({}^2S_{\frac{1}{2}} \right)$, to two excited electronic states $2p^6 3p \left({}^2P_{\frac{1}{2}} \right)$ and $\left({}^2P_{\frac{3}{2}} \right)$. The lamps emit monochromatic yellow light coming from the two transitions near 589nm as the excited states decay back down to the ground electronic state. The narrow 589nm emission can be filtered out to avoid light pollution in making astronomical observations.

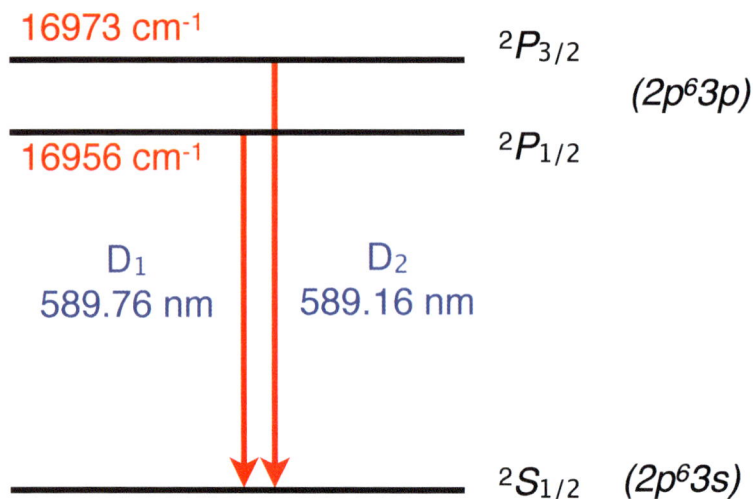

Figure 7.1: Electronic structure of the Na atom and transitions used in the sodium lamp.

7.3 Going Green with Mercury?

Fluorescent lamps, such as compact fluorescent lamps (CFLs), or energy saving bulbs, are designed as longer lasting, more energy efficient alternatives to incandescent bulbs. Instead of having a metal filament, all fluorescent lamps contain mercury, and are based on the physical chemistry of electronically excited mercury atoms. In order to understand how fluorescent lights work, we must first examine the energy level structure of the mercury atom.

In chapter 3, you learned that electrons in atoms and molecules can either be paired to give a singlet state, or unpaired to give a triplet state. The energy of the atom or molecule depends on this relative orientation of spins. Most stable molecules have a singlet as the ground state and a triplet as the first excited state. Transitions induced by absorption or emission of a photon between states with the same multiplicity are fully allowed in quantum mechanics, but transitions between states with different multiplicity are nominally spin *forbidden* (weak). In other words, a selection rule in quantum mechanics says that a photon cannot exert a torque on the spin of an electron in an atom to make a transition between a singlet and triplet state. Therefore, if an excited triplet state is populated through a process that does not obey the selection rules for light absorption/emission (i.e. electrical discharge, or high energy collison), the excited energy is stuck in the excited triplet state, and the atom or molecule cannot spontaneously give up its energy by emitting a photon and dropping back down to the ground singlet state. Excited singlet states of atoms typically have lifetimes of several hundred nanoseconds, and can spontaneously decay to the ground singlet state through the emission of a photon. The first excited triplet states of atoms can have much longer lifetimes on the order of seconds, since they cannot decay through photon emission. The low-lying excited triplet states of atoms and molecules are often called *metastable*, since they cannot spontaneously give up their excess energy through photon emission. Atoms and molecules in these metastable states can, however, drive chemical reactions through energy-transferring collisions with other nearby atoms and molecules.

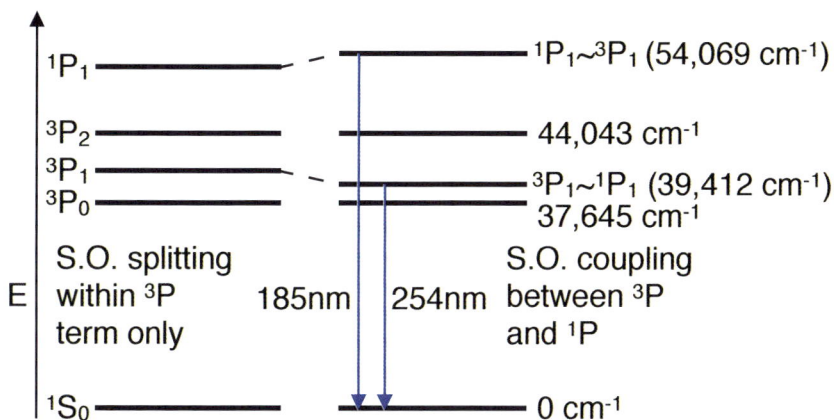

Figure 7.2: Energy level structure of the mercury atom.

Now that we understand the physical chemistry of triplet and singlet states, we can make a connection between the spin-orbit structure of the mercury atom and the mechanism of fluorescent lights by examining figure 7.2. Each CFL has a small amount of mercury inside a sealed, evacuated glass bulb. When you flip a light switch to turn on the bulb, an electrical discharge across the bulb excites some of the gaseous mercury atoms from the ground $6s^2$ $(^1S_0)$ state to the $6s6p$ (^3P) and (^1P) states. The $6s6p$ (^1P) state decays very quickly via strong UV emission at 185nm, but this light is not visible to the human eye.

Although the $6s6p$ $(^3P_1)$ $\rightarrow$ $6s^2$ $(^1S_0)$ transition is nominally spin forbidden, this is one of the strongest transitions in the mercury emission spectrum. How can this be? You already understand the spin-orbit splitting of the ^3P state into $^3P_{0,1,2}$ spin-orbit components (see left hand side of figure 7.2). However, there are off-diagonal matrix elements that couple and mix the $(^1P_1)$ and $(^3P_1)$ states (see right hand side of figure 7.2). For heavy atoms like Hg, L and S are not "good" quantum numbers. The "singlet" and "triplet" labels in the $^{(2S+1)}L_J$ term symbol are not exact. Instead, you have a mixture of singlet and triplet states that make the forbidden transition allowed. The selection rule for this off-diagonal spin-orbit mixing is $\Delta J = 0$. This is why only the $^1P_1$ and $^3P_1$ levels are coupled and mixed together, just like the two-level problem that we learned about in chapter 2. The coupled states can radiate at 185nm and 254nm due to this triplet~singlet spin-orbit mixing.

CFLs have a coating of phosphors on the inside of the bulbs. Once excited, the phosphors emit a broad spectrum of light in order to appear white to the eye and light up a room. CFLs use the Hg $6s6p$ $(^3P_1)$ $\rightarrow$ $6s^2$ $(^1S_0)$ emission at $254nm$ in the UV to make the phosphor coating on the inside wall of the bulb glow with white light. In addition, the metastable mercury atoms in the $^3P_2$ and $^3P_0$ states can transfer their excitation in collisions with the phosphor coating on the walls to generate the white phosphorescence.

Chapter 8

Lasers Are Quantum Devices

8.1 What is a Laser?

The word "laser" is an acronym that stands for "light amplification by stimulated emission of radiation". Laser light is different from sunlight or light from an incandescent bulb because laser light comes from stimulation of atoms or molecules in quantized excited states to emit photons, while normal light is spontaneously emitted all by itself, without any outside influence.

Laser light has special characteristics that ordinary light does not have. Laser light is intense (powerful) and directional (low divergence and small beam diameter). Laser light is also coherent (the light waves are all in phase with each other) and monochromatic (single frequency). Some lasers have tunable frequencies, and some emit light polarized in a particular direction in space. Some lasers emit pulses of light, while others provide a steady, continuous beam of light.

Lasers have many applications in physical chemistry, communications, medical diagnosis and treatment, drilling, cutting and welding, measuring distances, laser printing, reading and writing DVDs and reading barcodes. In fact, laser signals transmitted through optical fibers are the foundation of modern internet, cable television and telephone communications. A complete description of different types of lasers and their applications is beyond the scope of this book. We will simply make a connection between lasers, their applications and quantum mechanics. For more information on how different kinds of lasers work, and their many applications, review references [83] and [32].

8.2   How Do Lasers Work?

Lasers are inherently quantum mechanical in nature. In order to make a laser, a population inversion must be created in which a higher energy state has a larger population of atoms or molecules than a lower energy state. This population inversion can't be achieved under normal conditions when atoms and molecules are at thermal equilibrium. In other words, the energy population normally follows a Boltzmann distribution. For example, let's say we have a system with two energy levels. The probability ratio of atoms or molecules in the two states is,

$$\frac{P_2}{P_1} = e^{\frac{-E_2 + E_1}{k_B T}} \tag{8.1}$$

where P_i is the probability of finding an atom or molecule in state i=1 or 2, E_i is the energy of states i =1 and 2, $k_B = 1.38 \times 10^{-23}$ J/K is the Boltzmann constant, and T is the temperature of the system. According to the Boltzmann distribution, more atoms or molecules are in the lower energy state than the excited state. Therefore, any interacting photons have a higher probability of driving an absorption transition from the lower level to the excited level.

An inverted population would have more molecules in the excited state than in the lower state, and under these conditions a photon interacting with the atom or molecule is more likely to stimulate emission rather than absorption. As long as the population inversion holds, stimulated emission dominates. In order to create a population inversion, one must figure out a way to selectively excite atoms or molecules to a metastable (long-lived) energy state using light or electricity such that the rate of population of the metastable level exceeds the rate of decay.

Practical lasers use 3 or 4-level quantum systems. For a 3-level laser system (figure 8.1), imagine that most of the population starts off in the ground state (level 1), which acts as the lower level of the laser transition. Light or electrical excitation drives most of this population to a highly excited level (level 2) that is short-lived. The atoms or molecules in level 2 quickly decay to a lower metastable level (3), where the population builds up to create the inversion. The laser transition involves the stimulated emission from level 3 to level 1. In order for a 3-level laser system to work, you need an intense pulse of energy to depopulate level 1 and populate level 3.

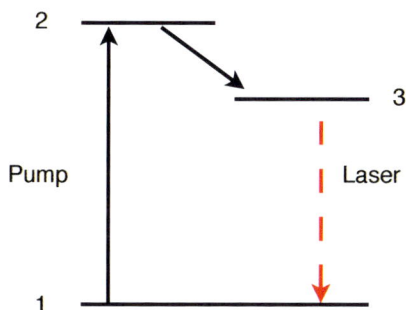

Figure 8.1: 3-level laser population inversion.

In 4-level lasers (figure 8.2), the first two steps are the same as in the 3-level system. Excitation depopulates the ground state, and populates a short-lived excited state that decays to a metastable state to create the population inversion. The key difference is that the laser transition takes place between the metastable level and a 4th level that lies above the ground state. This way, you do not need a large pulse of energy to depopulate the ground state to sustain the population inversion, since the ground state is not the lower level of the laser transition.

8.3 Lasers in Physics and Chemistry

8.3.1 Spectroscopy and Dynamics

Tunable lasers that have a narrow linewidth are by far the most important tool we have for learning about the quantized energy level structure of atoms and molecules. Laser light can selectively drive electronic transitions in atoms and rovibronic (rotational, vibrational and electronic) transitions between molecular energy levels. If we can detect the absorption of light or photons emitted

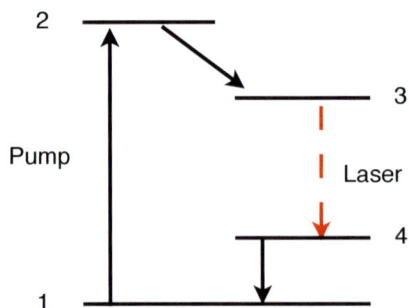

Figure 8.2: 4-level laser population inversion.

in such transitions, we can map out the energy level structure of the atom or molecule. This is why tunable lasers are used in spectroscopy (the study of the interaction of light with matter) to record a spectrum (a plot of atomic or molecular absorption or emission intensity as a function of light frequency) [33–35]. Lasers are one of the most important and essential tools in experimental physical chemistry research. They can be used to determine the chemical and physical composition, bonding, structure and properties of matter [10, 86, 87].

The most famous, founding experiment of quantum mechanics is the Stern-Gerlach experiment [28]. In the experiment, Otto Stern and Walther Gerlach proved the existence of space quantization using an atomic beam [12, 55, 63, 70] of silver atoms that was split with a magnetic field. Their groundbreaking work later led to the inclusion of electron spin in quantum theory. Although no lasers were used in the Stern-Gerlach experiment, Stern's pioneering work on atomic beams led to experiments with molecular beams.

In a pulsed molecular beam experiment [12, 55, 63, 70], a pulse valve sends a supersonic, free-jet expansion of gas molecules into a vacuum chamber that can have a pressure billions of times lower than atmospheric pressure. The conical expansion spontaneously cools and can be skimmed to form a cold beam of molecules with rotational temperatures close to absolute zero. This experimental technique effectively cools molecules down to their lowest energy states so that state-selected laser spectroscopy can be performed in order to track the energy flow from one quantum state to the next in an attempt to simulate a chemical reaction. This technique opens up the possibility for laser initiation, control and elucidation of the mechanism and dynamics of chemical reactions.

Now that we understand the basic physics of lasers and the fundamental relevance to chemistry research, we can learn about a few important applications of lasers and quantum mechanics in the next two chapters.

Chapter 9

Quantum Sensing

Spectroscopy [87] allows us to bridge the world of light (electro-magnetic radiation) and matter (atoms and molecules). We have seen that this light-matter interaction is key for laser probes of atomic and molecular structure, chemical bonding and dynamics of chemical reactions.

Spectroscopy can be divided into groups, based on the types of transitions excited by different frequencies of electromagnetic radiation. First, long wavelength radiation (in the microwave to mm-wave range) can induce transitions between quantized rotational energy levels in molecules. Second, infrared light activates vibrations of molecules and causes bonds to symmetrically or asymmetrically stretch, bend or undergo torsional motion. Many absorption spectroscopy experiments are carried out in the infrared, so that molecules can be identified based on their characteristic vibrational frequencies that depend on molecular structure. Third, light in the visible through UV range drives electronic transitions, where both the energy and geometry of a molecule can be profoundly different from one electronic state to the next. Finally, high-energy UV light can either break bonds in molecules to probe the bond-rupturing dynamics that drive chemical reactions, or photoionize molecules to form charged cations. In this chapter, we will see that fundamental research involving spectroscopy has led to transformative improvements in the fields of medicine and sensing.

9.1 Infrared Absorption Spectroscopy

9.1.1 Traditional IR Absorption Spectroscopy

The infrared region of the electromagnetic spectrum spans about 100 - 5000 cm^{-1} = 1.2 - 60 kJ/mol = .3 - 15 kcal/mol. These energies are sufficient for excitation of molecular vibrations. Rotational excitations are smaller in energy, so light that can excite vibrations will usually simultaneously excite rotations. This results in a spectrum with fine rotational structure on top of coarse vibrational structure. In order to record a spectrum, the infrared light is passed through a sample, and intensity is measured as a function of laser frequency.

A traditional absorption spectroscopy experiment starts by splitting an infrared laser beam in two. One beam passes through a sample cell that contains a mixture with the target absorbing molecules, and the other beam passes through a reference cell that contains the same mixture without the target absorbers. If the laser frequency is tuned into resonance with a vibrational transition in the target molecule, some fraction of the molecules in the sample will absorb the resonant light. One can then measure how much of the light has been absorbed in order to measure the concentration of the sample and identify the absorbing species based on its characteristic spectrum. Two of the most important quantities in any absorption spectroscopy experiment are the intensity of the reference beam, I_0, and the intensity of the transmitted beam after it has passed through the sample, I_t. Measurements are made by determining absorption, which is defined as the difference, $I_0 - I_t$. If the light is absorbed by a large number of molecules, $I_0 - I_t$ will be a measurable quantity. The detection limit is reached for small numbers of absorbers as $I_0 - I_t$ approaches zero. Therefore, the main criterion for the feasibility of a traditional absorption measurement is that the signal-to-noise ratio must be large enough to distinguish between I_0 and I_t.

The main problem with traditional absorption spectroscopy is that the goal of most applications is to detect target molecules that are present in trace amounts in a large volume of air or background gas. These trace amounts result in small absorption, $(I_0 - I_t \approx 0)$. One way to increase the number of absorbers interacting with the laser is to increase the path length of the laser beam through the

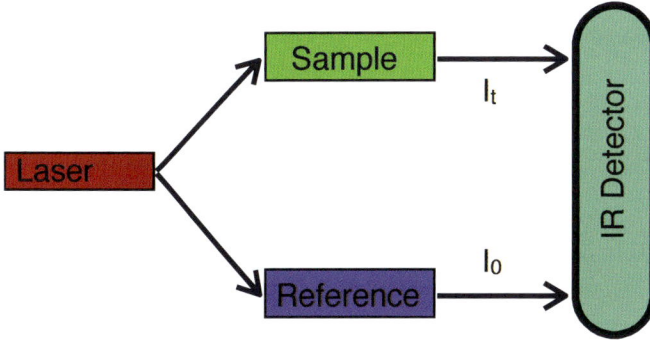

Figure 9.1: Traditional absorption spectroscopy experiment.

sample. One method of increasing the path length is to make a large spectrometer that can contain the entire volume of gas to be probed. However, this solution is costly and not practical for applications in remote sensing, analysis of high-purity gases in the semiconductor industry, combustion research, atmospheric pollution monitoring and control, detection of chemical warfare agents on the battlefield and detection of illegal drugs or explosives in baggage-handling areas of airports. In these cases, inexpensive, compact and portable instruments are required. An alternative solution to the problem of increasing the laser path length through a sample is to pass the laser beam through the sample multiple times, before measuring absorption using a White cell or Herriott cell, where an optical cavity is formed by two mirrors that reflect the laser beam back and forth between the mirrors, until the laser exits the cavity for detection.

9.1.2 Cavity Ringdown Spectroscopy

Cavity Ringdown Spectroscopy (CRDS) [86] is an absorption technique for sensitive detection of trace molecules in gaseous samples. CRDS is more sensitive than traditional absorption spectroscopy, because it does not rely on measuring absorption as a small change in the intensity of light, $I_0 - I_t$, against a large background, I_0.

Instead, CRDS monitors the exponential decay of light in an optical cavity, leading to much higher detection limits compared to traditional absorption spectroscopy.

Two highly reflective mirrors form the CRDS optical cavity. A short pulse of light enters the cavity and reflects back and forth between the mirrors, effectively increasing the path length. If the cavity is evacuated, one mechanism for the loss of light is transmittance through the mirrors. The amount of light lost through each pass is a function of the reflectivity of the mirrors. The intensity of light decays exponentially with time. In other words, the decay resembles a ring-down pattern, much like the decay of sound from a ringing bell. The measured quantity in the case of an evacuated cell is called the *cavity ringdown time*, τ_{empty}, for the exponential decay of light intensity. If an absorbing sample is present inside the cavity, the ringdown time is written as τ, where $\tau < \tau_{empty}$.

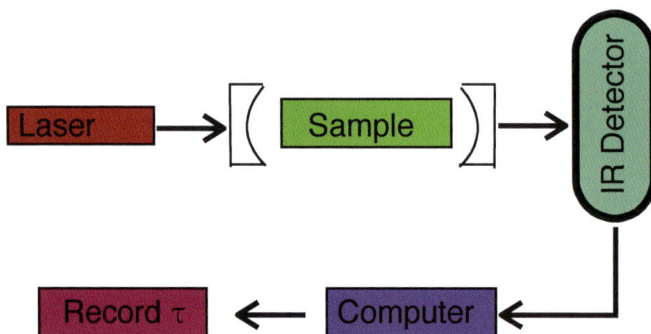

Figure 9.2: CRDS experiment block diagram.

The exponential decay time can be mathematically related to the concentration of sample gas in the cavity using the Beer-Lambert Law,

$$I_t = I_0 e^{-\epsilon C L} \tag{9.1}$$

where C is the sample concentration in mol/cm^3, L is the path length of the absorbing sample in cm and ϵ is the molar absorption coefficeint in cm^2/mol. ϵ is a constant for a given molecule that reflects the probability of an absorption transition driven by a photon.

Equation 9.1 can be rearranged to give:

$$\frac{I_0}{I_t} = e^{\epsilon C L} \tag{9.2}$$

$$ln\frac{I_0}{I_t} = \epsilon C L \tag{9.3}$$

where Beer's Law is traditionally expressed as,

$$A = \epsilon C L \tag{9.4}$$

with A, the absorbance of the sample defined as,

$$A = ln\frac{I_0}{I_t} \tag{9.5}$$

Using the Avogadro constant $(6.02 \times 10^{23} \text{ mol}^{-1})$ to convert units from moles to molecules, equation 9.4 can also be expressed as,

$$A = \sigma^2 N L \tag{9.6}$$

where σ^2 is an absorption coefficient with units of $\text{cm}^2/\text{molecule}$, N is the concentration of the sample in $\text{molecules}/\text{cm}^3$, and L is the path length through the sample in cm.

The relationship between equation 9.6 and the cavity ringdown times is [86]:

$$\frac{A}{L} = \sigma^2 N = \frac{1}{c}\left(\frac{1}{\tau} - \frac{1}{\tau_{empty}}\right) \tag{9.7}$$

where c is the speed of light. Using this equation 9.7, measurement of the cavity ringdown times, τ and τ_{empty}, gives the absorbance, A, and number of absorbing molecules, N, in the sample.

The practical applications of CRDS are nearly limitless. Imagine if the military could send a small, portable CRDS device on a remote controlled vehicle into the battlefield to detect chemical warfare agents to ensure safety prior to deploying troops in the area. This technology is all based on the resonance between the quantized energy levels of molecules. Next, we will give a brief review of some other applications of quantum mechanics to chemical detection and disease diagnosis.

9.2 Applications

9.2.1 Analysis of Human Breath

Future disease diagnosis will be markedly different from the way it is performed today. Advances in technology and novel scientific ideas are enabling noninvasive alternatives to traditional diagnostic techniques. Breath analysis is one next-generation technique for detecting and monitoring human illness. Analysis of exhaled breath has applications in diagnostics of airway inflammation, pulmonary diseases, oxidative stress, liver and kidney dysfunction, and metabolism. On the surface, breath analysis appears to be a conceptually simple tool for disease detection, but several key aspects of the problem must first be considered before it can be implemented in the field of medicine. What instruments and techniques are suitable for analyzing exhaled breath, and what will these instruments monitor as indicators of disease? Development of sensor technology is the initial step in answering these key biological questions relevant to noninvasive disease diagnosis. In exhaled breath, one can look for biomarkers specific to a particular disease. The biomarker may be a protein or small molecule. Selective and highly sensitive techniques, such as CRDS, are necessary in order to relate the species exhaled in breath to specific diseases. The goal is to develop advanced technology for the identification and monitoring of the disease biomarkers. It is clear that an interdisciplinary approach, which combines cutting-edge sensing technology with advanced knowledge of biological systems, will play a key role in the development of sensors for biomarker detection. The new generation of ultra-sensitive, selective bio-sensing systems will emerge at the interface between physical chemistry and biology. The fundamental operation of the advanced sensing technology relies heavily on quantum mechanics, where resonances between the laser light from the sensor and the quantized energy levels of target biomarker molecules are the keys to selective detection.

9.2.2 Raman Spectroscopy in Airport Security

Raman spectroscopy [33] involves the observation of light that has been scattered by molecules in a sample, rather than the light transmitted through a sample. If laser light with frequency ν_0 is directed through a sample, most of the light will pass through the sample with the same frequency. A small percentage of that incident light will scatter isotropically with the same frequency, ν_0. This is called Rayleigh scattering. Rayleigh scattering explains why the sky is blue. Light from the sky on a clear day is just light that has been scattered off of molecules in the atmosphere. The blue light is scattered more effectively than red light, so the sky appears blue. Raman scattering is a different phenomenon. A very small percentage of the scattered light will have a frequency that is different from ν_0. This Raman scattered light has frequency ν_i, where $\Delta E = h|\nu_0 - \nu_i|$. Raman scattering is currently used in the detection of explosives and illegal drugs in airport security.

9.2.3 Green Fluorescent Protein

The 2008 Nobel Prize in Chemistry was awarded to Martin Chalfie, Osamu Shimomura and Roger Tsien for the discovery and development of the green fluorescent protein, GFP [71]. What is GFP, why was GFP research awarded the Nobel Prize, and how is all this related to quantum mechanics?

All proteins are made of amino acids, linked in chains. Some of the amino acids in GFP form a chromophore, a chemical group that can absorb light, and spontaneously fluoresce. The chromophore in GFP is activated by UV light, and fluoresces in the green. GFP was discovered in the Pacific Ocean jellyfish, Aequorea victoria.

Shimomura, Chalfie and Tsien were the pioneers in using the fluorescence from GFP as a beacon for activity within cells. GFP has been used to map out the way cancer tumors grow and form new blood vessels and to track the development of Alzheimer's disease. At the fundamental level, the GFP fluorescence is due to the quantized energy level structure of the molecule.

9.2.4 Lasers in Combustion Research

Laser-induced fluorescence (LIF), Raman spectroscopy and Rayleigh scattering have all been used to study combustion in an effort to improve efficiency and reduce pollutant emissions in engines. This research in the fundamental gas-phase physical chemistry of combustion is driven by the increasing demand for clean energy and desire for affordable, efficient energy conversion, long-term energy storage, and controlled energy release for sustainable use. The global objective is an energy efficient, environmentally benign future for our ever-changing global energy infrastructure.

Combustion science is challenging because of the turbulence of chemically reacting systems. The combustors in energy generation systems have a very large scale, and the combustion propagation and mixing in internal combustion engines is complex. Laser-based diagnostic tools can be used to image the turbulent flow of a fuel as it burns in oxygen. The combustion of hydrocarbon fuels always forms highly reactive intermediates such as HOCO, NH, NH_2, OH and CH_3. Laser spectroscopy can be used to photolyze parent precursor molecules to create and probe the properties of these open-shell (molecules with unpaired electrons) combustion intermediates that mediate the combustion reactions. The basic science goal is to take advantage of our knowledge of the quantum mechanics of molecules to control matter and energy related to combustion at the microscopic level.

9.2.5   Magnetic Resonance Imaging

You have learned that many nuclei have non-zero spin angular momenta that give rise to magnetic moments. One powerful technique used to detect the magnetic moments of nuclei is nuclear magnetic resonance (NMR). NMR is widely used in chemistry research to determine the structures of molecules. In the medical field, the technique is called magnetic resonance imaging (MRI).

H-atom nuclei in an external, strong magnetic field can be aligned either parallel or anti-parallel with the field, giving rise to two different quantized energies. Radio-frequency (rf) electromagnetic radiation can be absorbed by the sample and cause the nuclear spins to flip, driving a transition between the two levels. This resonant absorption of energy is detected by the NMR or MRI instrument. The difference in energy between the two spin states is dependent on the chemical environment of each nucleus. Therefore, by probing a molecular sample containing nuclei with non-zero spin (i.e. H or ^{13}C) in a strong magnetic field with rf pulses, one can map out the chemical environment of the sample.

The MRI technique takes advantage of the fact that our bodies are largely made of water, H_2O. Water contains atomic hydrogen nuclei (protons). In MRI, a patient is placed inside a strong magnetic field, and the soft tissues to be probed are hit with pulses of rf electromagnetic radiation. The nuclei are excited by the rf pulses, and they relax back down to a lower energy state. This relaxation time is highly dependent on the type of tissue being probed, and the image can be built from analysis of these relaxation times.

Magnetic resonance imaging (MRI) has revolutionized the visualization of soft tissue (muscle, internal organs) in medicine. MRI is complementary to X-ray techniques that are primarily used for imaging hard structures like bone, or teeth. An MRI image simply reflects the concentrations of protons in tissue.

One thing that makes MRI especially useful is the fact that MRI images of water protons are different in healthy vs. diseased tissue. This enables the use of MRI as a diagnostic tool in medicine. Contrast agents are paramagnetic compounds that are distributed differently in healthy vs. diseased tissue. These compounds can be injected into the body and used to enhance image contrast for disease diagnosis.

Functional MRI (fMRI) is widely used to map out blood flow in the brain, because fMRI is sensitive to the magnetic properties of oxygenated vs. deoxygenated hemoglobin, the oxygen transporting protein of red blood cells. There is greater blood flow in regions of the brain that are active compared to inactive regions of the brain. Therefore, the flow of blood in the brain can be correlated with the mental activities of a patient. This has revolutionized the field of neuroscience. Both NMR and MRI are entirely based on the quantum mechanics of nuclear spin.

Chapter 10

The Chemistry of Global Climate Change

10.1   Atmospheric Photochemistry

Atmospheric pollution and global climate change are governed by the rules of quantum mechanics. This is because many reactions in our atmosphere are initiated and driven by sunlight. Molecules in the atmosphere absorb sunlight at specific frequencies, in accord with the allowed transitions between the quantized energy levels of the molecules. This can result in rotational, vibrational or electronic excitation. Photon absorption might also result in the rupture of chemical bonds to form atomic or molecular fragments, or ionization to form cations. Since molecules in their excited electronic states are much more chemically active than the same molecules in their ground electronic states, the molecular absorption of sunlight is key to producing chemically active species in photoexcited states that can take part in reactive collisions with other species. After we review some fundamentals of real-world atmospheric chemistry, we will discuss the possible fates of these photoexcited molecules. We will then examine the cutting-edge laser-based techniques that are used to study atmospheric photochemistry in the research laboratory.

10.2   Oxygen and Ozone Photochemistry

More than 20% of the gas in our atmosphere today is O_2, but this large percentage of molecular oxygen was not present when our planet formed. The production of atmospheric O_2 can't be explained through inorganic photochemistry alone [93]. Venus and Mars, our two neighboring planets, have a very small percentage of O_2. Where did the large percentage of O_2 on Earth come from? The answer is photosynthesis. The oxygen in our atmosphere has been formed nearly entirely from the biological activity of plants.

Now that we know where the oxygen in our atmosphere comes from, let's discuss the importance of the oxygen to life on earth. Of course, you know that we need oxygen for respiration. You may not know that O_2 also acts as a chemical filter for harmful UV light, because oxygen has absorption bands for short wavelength UV light below 230 nm. However, since the DNA in our bodies can be damaged by UV light with wavelengths shorter than 290 nm, we need some other protection for wavelengths from 230 - 290 nm. This additional protection comes from ozone, O_3. We owe our life on Earth's surface to O_3 and O_2. Without these two chemicals, we would not survive on dry land.

10.2.1   Production of Stratospheric Ozone

A major focus of research in atmospheric chemistry is on reactions related to the production and destruction of ozone and its photochemical precursors, like O_2. The Chapman Mechanism (1930) describes the process for the formation of the ozone layer in the stratosphere, which is a region of the atmosphere about 25-30 km above sea-level.

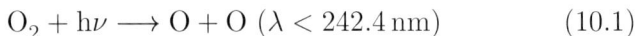

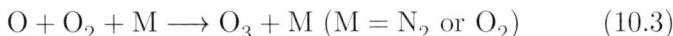

$$O_2 + h\nu \longrightarrow O + O \ (\lambda < 242.4\,\text{nm}) \tag{10.1}$$

$$O_3 + h\nu \longrightarrow O_2 + O \ (\lambda < 1180\,\text{nm}) \tag{10.2}$$

$$O + O_2 + M \longrightarrow O_3 + M \ (M = N_2 \text{ or } O_2) \tag{10.3}$$

$$O + O_3 \longrightarrow O_2 + O_2 \tag{10.4}$$

At very high altitudes, there is plenty of short wavelength UV light to drive the photodissociation of O_2 (reaction 10.1). Lower altitudes contain a much higher concentration of O_2, but the short wavelength UV light cannot penetrate far through the atmosphere, because it has been absorbed by molecular oxygen. Ozone exists in a layer in between these two regions.

The fate of most ozone in the stratosphere is dissociation by sunlight. This natural photodissociation is a key part of the global cycle of ozone production and destruction. It is well known that this natural cycle of light absorption and photodissociation involving oxygen and ozone in the stratosphere is what protects organisms living on Earth's surface from harmful UV wavelengths. The quantum interaction between sunlight and molecules has control over this natural cycle of oxygen/ozone production and destruction.

10.2.2   Depletion of Stratospheric Ozone

Years ago, molecules called chlorofluorocarbons (CFCs) were used in refrigeration technology. These molecules are normally very inert, and the only way for them to become chemically active is by rising up from the ground level into the stratosphere. One particular CFC is dichlorodifluoromethane, CF_2Cl_2, otherwise known as (CFC-12). Once CF_2Cl_2 rises to the stratosphere, it can be photodissociated by UV light:

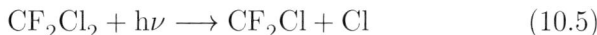

$$CF_2Cl_2 + h\nu \longrightarrow CF_2Cl + Cl \qquad (10.5)$$

This reaction is particularly bad for the ozone layer, because the product Cl atoms destory ozone in a catalytic cycle:

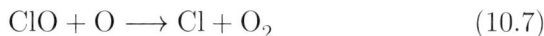

$$Cl + O_3 \longrightarrow O_2 + ClO \qquad (10.6)$$
$$ClO + O \longrightarrow Cl + O_2 \qquad (10.7)$$

You might have read about the ozone hole that exists over the South Pole above Antarctica. The hole in the ozone layer is located above the South Pole, because this is a very cold region of the atmosphere where polar stratospheric clouds exist. The following reactions take place in polar stratospheric clouds:

$$ClO + NO \longrightarrow NO_2 + Cl \qquad (10.8)$$
$$ClO + NO_2 \longrightarrow NO_3Cl \qquad (10.9)$$

A product of these reactions, chlorine nitrate, condenses in these cold polar stratospheric clouds in the cold winter months. When the clouds warm up in the spring, NO_3Cl is released into the gas-phase and photodissociated to produce Cl. The catalytic cycle of ozone destruction by Cl (reaction 10.6) then continues in this local region of the atmosphere. Again, the driving force behind the CFC depletion of stratospheric ozone begins with the photodissociation of molecules like CF_2Cl_2. Quantum mechanics plays a vital role.

10.2.3 Production of Tropospheric Ozone

While ozone in the stratosphere is vital to life on Earth, ozone in the troposphere is hazardous to life. Human production of nitric oxides (NO_x) is the main source of this *toxic* ozone. Specifically, tropospheric ozone is produced by photodissociation of NO_2 and consumed in reactions with NO.

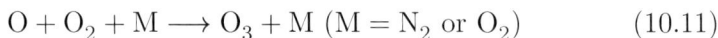

$$NO_2 \xrightarrow{h\nu} NO + O, (\lambda < 400 \text{ nm}) \tag{10.10}$$

$$O + O_2 + M \longrightarrow O_3 + M \ (M = N_2 \text{ or } O_2) \tag{10.11}$$

$$NO + O_3 \longrightarrow NO_2 + O_2 \tag{10.12}$$

Reaction 10.11 is the main reaction that produces ozone in the atmosphere. The relationship between NO_2 and NO described by reactions 10.10 and 10.12 is the determining factor in the amount of O_3 produced.

In a related set of coupled reactions, oxidation of CO is initiated by atmospheric OH radicals to produce tropospheric ozone when NO_x is also present.

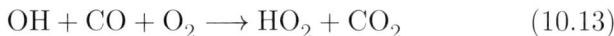

$$OH + CO + O_2 \longrightarrow HO_2 + CO_2 \tag{10.13}$$

$$HO_2 + NO \longrightarrow OH + NO_2 \tag{10.14}$$

The generation of NO_2 in reaction 10.14 again drives the production of O_3 as in reactions 10.10 and 10.11.

10.3 Global Warming

In reaction 10.13, it can be seen that the oxides of carbon play a key role in tropospheric ozone photochemistry. Of course, carbon dioxide is also a major greenhouse gas, and CO_2 emissions from burning of fossil fuels are causing global climate change in both the upper and lower atmosphere.

Global warming is a result of the quantum interaction between sunlight and atmospheric molecules. The process of global warming starts with sunlight warming the surface of the earth. The earth reradiates this energy into the atmosphere as infrared light. This is why atmospheric molecules exist in a natural bath of mid-infrared radiation. Molecules such as CO_2 absorb specific frequencies of this infrared light according to their quantized energy level structure and quantum selection rules. The absorption and vibrational excitation is the cause of global warming.

10.4 OH Radicals in the Troposphere

About 90% of the mass of the atmosphere is contained in the troposphere. It turns out that hydroxyl radicals, OH, dominate tropospheric chemistry. OH can oxidize molecules such as hydrocarbons, H_2 and CO to form H_2O and CO_2. In this way, hydroxyl radicals act as scavengers for many different kinds of pollutants in the troposphere. The oxidation reactions are like combustion reactions at low temperatures, because they start with a molecular fuel, and end with water and carbon dioxide as products. All of these reactions involve photochemistry.

The main mechanism for production of OH in the troposphere [45] is the photodissociation of ozone to generate oxygen atoms in their excited state, O (^1D). The majority of the O (^1D) atoms are quenched back to the ground state, O (^3P), by collisions, but some of them can react with water to form OH:

$$O_3 + h\nu \longrightarrow O_2 + O(^1D) \ (\lambda \leq 320\,\text{nm}) \qquad (10.15)$$

$$O(^1D) + M \longrightarrow O(^3P) + M \ (M = N_2 \text{ or } O_2) \qquad (10.16)$$

$$O(^1D) + H_2O \longrightarrow 2\,OH \qquad (10.17)$$

Another mechanism for the production of OH involves the electronic excitation of NO_2 by visible light to become NO_2^*. The energy provided by the electronic excitation provides nitrogen dioxide with sufficient energy to overcome the barrier to reaction with water and form OH:

$$NO_2 + h\nu \longrightarrow NO_2^* \ (\lambda > 420\,\text{nm}) \tag{10.18}$$

$$NO_2^* + M \longrightarrow NO_2 + M \ (M = N_2 \, , \ O_2 \text{ or } H_2O) \tag{10.19}$$

$$NO_2^* + H_2O \longrightarrow OH + HONO \tag{10.20}$$

The NO_2^* species has a long fluorescence lifetime (40-60 μs), so it will survive long enough to collide and react with H_2O.

10.5 Photochemical Pollutants

Due to absorption by O_3 and O_2, only wavelengths $\lambda > 290$ nm make it down to the troposphere. At these wavelengths, two of the most important photochemically active molecules are O_3 and NO_2. In the troposphere, ozone and nitrogen dioxide are pollutants that can cause chemical reactions associated with bad smells, brown haze, and reduced visibility. These species and their reaction partners can damage rubber and vegetation. One molecule produced through NO_2 photochemistry is peroxyacetyl nitrate (PAN). PAN ($CH_3COOONO_2$) is toxic to plants, and can irritate the eyes and respiratory system.

10.6 Molecular Disposal of Energy

We have learned that the quantized molecular absorption of light provides the energy for driving atmospheric chemical reactions. Molecules tend to spontaneously lose energy that has been absorbed in order to return to their most stable, ground electronic state. One important question to be answered is, after a molecule has absorbed energy from a photon of light, what are the available pathways for disposal of this energy?

10.6.1 Radiative and Non-Radiative Processes

Radiative processes for the spontaneous disposal of electronic energy through emission of light include fluorescence and phosphorescence. Fluorescence involves radiation to a lower state of the same multiplicity as the upper electronic state. Phosphorescence is radiation to a lower state of different multiplicity than that of the upper electronic state. One type of non-radiative de-excitation process is called InterSystem Crossing (ISC) [4]. In ISC, energy flows from vibrational levels of one electronic state to another electronic state of different spin multiplicity via radiationless transitions.

Figure 10.1 presents the processes of fluorescence, ISC and phosphorescence for a molecule. First, a photon of light can excite the molecule from the ground singlet electronic state, S_0, to the first excited singlet state, S_1. Direct photon excitation from S_0 to T_1, the first excited triplet state, is spin-forbidden. Once S_1 is populated, the molecule can give up its electronically excited energy via fluorescence back down to S_0, with a fluorescence rate constant, k_f.

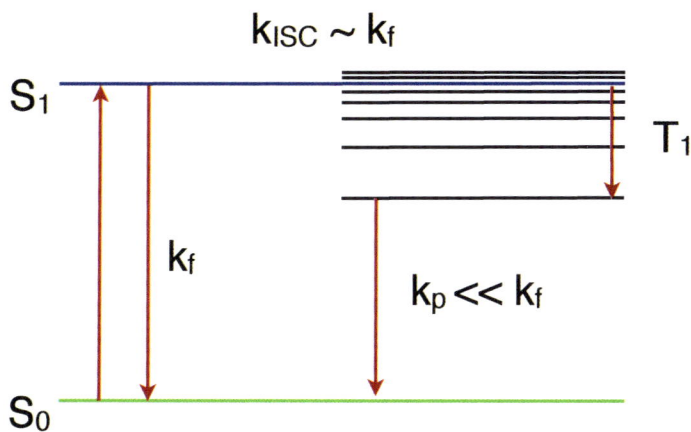

Figure 10.1: InterSystem Crossing, ISC.

In addition, ISC can take place between the excited vibrational level in S_1 and highly excited vibrational levels of T_1 that are near-degenerate (close in energy). The rate constant for the ISC that

109

competes with fluorescence is k_{ISC}. Once the highly excited vibrational levels of T_1 are populated, collisional vibrational relaxation can take place down to the ground vibrational level of T_1. The ground vibrational level of T_1 is metastable (long-lived), due to the forbidden transition back to the ground electronic state, S_0. Therefore, the rate constant of phosphorescence, k_p, is much smaller than k_f. The rate of ISC can be described using Fermi's Golden Rule [81]:

$$k_{ISC} = \frac{2\pi}{\hbar}\rho(E)|\langle S_1|\hat{H}^{S.O.}|T_1\rangle|^2 \qquad (10.21)$$

where $\rho(E)$ is the density of T_1 levels near the singlet S_1 level, and E is the excess energy above the ground vibrational level of T_1. The rate of ISC is controlled by the squares of spin-orbit matrix elements between S_1 and T_1. For deeper insight into the theory and experimental observation of ISC in molecules such as acetylene, C_2H_2, read the article by Bittinger, Virgo and Field [4].

10.6.2 Laser Photodissociation

If an absorbed photon has enough energy, it can cause a molecule to fragment. Bond-breaking due to the action of light absorption is called photodissociation. In this chapter, we have seen examples of the photodissociation of oxygen, nitrogen dioxide and ozone, leading to secondary reactions that drive the chemistry of the atmosphere. Let's examine the photodissociation of a triatomic molecule, ABC:

$$ABC + h\nu \rightarrow ABC^* \rightarrow AB + C \qquad (10.22)$$

It is often the case that the energy of the dissociating photon, $h\nu$, is greater than the energy of the bond being broken, $D_0(B-C)$. Since energy is conserved, the available excess energy must be transformed into the translational or internal energy of the photofragments. There are several important questions to address with regard to this photodissociation reaction:

- What is the quantum mechanism for excitation and bond rupture?

- What is the lifetime and structure of the excited ABC^* complex?

- Why are photofragments AB and C formed, and is the fragmentation pathway wavelength dependent?

- What is the relationship between the properties of the absorbed light and the trajectories of the photofragments?

- How is the excess photon energy partitioned into the translational kinetic energy and internal, quantized energy states of the photofragments?

These are questions that laser spectroscopy can answer. One possible mechanism of photodissociation is *direct dissociation* [67], where the excitation is to a *dissociative*, or *repulsive* electronic state, and the molecule falls apart on this dissociative potential energy surface. Tunable pulsed lasers are ideal for laboratory photodissociation studies for two reasons. First, they output a high enough power to dissociate a large fraction of a sample of molecules. Second, they have high spectral resolution to excite individual quantized energy levels that will lead to dissociation. Using lasers, one can map out the quantum mechanical structure of photodissociation.

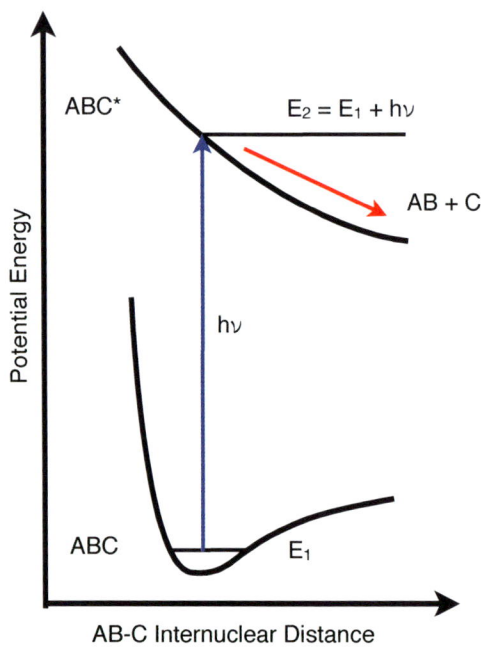

Figure 10.2: Mechanism of direct dissociation on a single excited repulsive potential energy curve.

10.6.3 Laser Photoionization

Another route to disposal of absorbed photon energy is photoionization, the ejection of an electron from an atom or molecule. Knowledge of photoionization is important in atmospheric chemistry, since many of the ions and electrons present in the Earth's atmosphere are formed as a result of photoionization. In the laboratory, it is often the case that one photon of light from a visible or UV laser will not have enough energy to ionize a light atom or molecule in one step. Therefore, in laser photoionization studies, multiphoton ionization (MPI) is used [38, 39]. With MPI, if the laser power is high enough, a molecule can absorb multiple photons to the ionization continuum where an electron is ejected.

In the situation where the absorbed photons are in resonance with a bound, quantized energy level of the atom or molecule, the process is called resonance-enhanced multiphoton ionization (REMPI). An example with the hydrogen atom is given here [95].

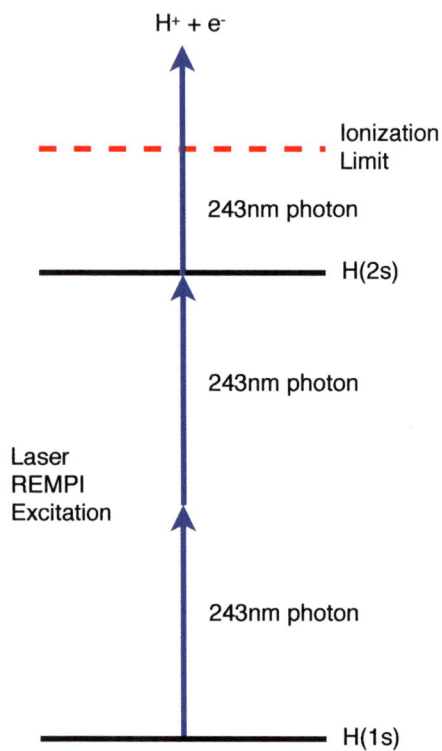

Figure 10.3: MPI of the hydrogen atom.

10.7 Laser Velocity Map Imaging (VMI)

Laser Velocity Map Imaging (VMI) [14–16, 41, 82, 94] is a cutting-edge fundamental experimental research technique that has transformed the scientific community's understanding of photochemistry. VMI uses the tools of laser photodissociation and photoionization to reveal the pathways for bond rupture and the dynamics of energy flow into the rovibronic internal states of photofragments.

VMI research is transformative, because it simultaneously enables advanced discovery, increased knowledge, deepened understanding and the merging of several scientific spheres. The field of chemical dynamics [96] benefits from VMI research through an understanding of how one can experimentally observe, and possibly control, reaction pathways. This is one of the most challenging, yet fundamental problems in chemistry. The models for photochemical energy flow developed from these experiments are used by photochemists to understand the relationship between dissociation and other routes for the loss of electronic excitation such as ionization, luminescence (fluorescence and phosphorescence), bond-breaking isomerization, *cis-trans* isomerization and collisional quenching. Atmospheric chemists gain insights from VMI to upgrade their models [53, 62] of mixing and transport, pollution control and photodissociation cross sections. Building upon the results of this fundamental research, environmental scientists can model the complexities of intermolecular and intramolecular energy transfer, and apply the models to understand phenomena such as chemical dynamics at the gas-liquid interface where the atmosphere meets the ocean surface. Photodissociation is the key chemical process that binds all of these phenomena. In photodissociation, when new products are formed through fragmentation, they may go on to participate in secondary thermal reactions that would not have occurred in the absence of light. The resulting photofragments act as a sink for starting chemicals, and drive complex chemical changes as they carry different amounts of translational and rovibronic internal energy.

Most chemical reactions in the atmosphere are initiated and driven by photochemistry. In atmospheric photodissociation reactions, a portion of the dissociation photons have an energy significantly greater than the dissociation energy of the chemical bond

115

which is being broken. The main fundamental scientific problem to understand is how this excess energy in a photodissociation reaction is partitioned between the resultant photofragments. The partitioning of energy is one of the major factors that mediates all photochemistry. Addressing this problem necessarily requires innovation through boundary-crossing research that can move in profoundly new, promising directions.

10.7.1 Molecular Beam Laser Apparatus

The VMI experiment takes place inside a differentially-pumped pulsed molecular beam [12, 55, 63, 70] vacuum apparatus (figure 10.4) that can achieve pressures that are ten billion times lower than atmospheric pressure. The first chamber in the vacuum apparatus, the *Source*, houses a supersonic pulse valve containing the target gas molecules. The pulsed free-jet expansion of gas molecules is skimmed to form a collimated beam of molecules in their lowest energy electronic and vibrational states that travels into a second chamber where the photochemical reaction takes place.

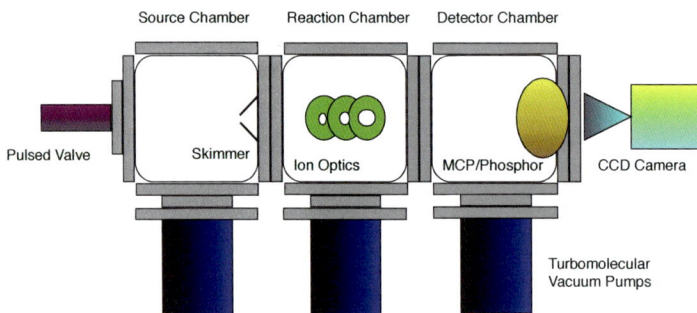

Figure 10.4: Virgo design for a differentially pumped molecular beam apparatus with ion optics and photofragment detector for velocity map imaging.

In the *Reaction* chamber, the molecular beam is crossed with a UV dissociation laser of known polarization. The resulting photofragments fly outward from the center-of-mass with measurable velocities. Soon after the photolysis laser pulse, another laser pulse selectively ionizes one of the photoproduct species. Photoionization is a key step for detection of the photofragments. Neutral fragments are difficult to manipulate and detect, unless they reveal themselves through fluorescence. However, cations are readily detectable, and their motion can be manipulated through the use of electric fields.

10.7.2   Ion Optics, MCP Detection and CCD Imaging

The resulting cations must be focused into the *Detector* chamber in order to measure product translational kinetic energy, angular distribution and branching of energy between products [54]. A set of electrostatic lenses, called ion optics [18, 21, 25, 46], is used to perform this task. Ion optics are similar to lenses that focus visible light, but ion optics use electric fields to focus charged particles. The ion optics accelerate product cations onto a position-sensitive MCP/phosphor detector. The MCP/phosphor detector is very similar to the detector used in mass spectrometers that can be found in any university chemistry department, or in many airport security stations. The difference is that mass spectrometers detect the mass-to-charge ratio of ions, while MCP/phosphor detectors are sensitive to the positions of charged particles that hit the detector, and the phosphor screen lights up when the detector is hit by a charged particle to indicate its location. A sensitive CCD camera then takes a high-resolution image of the phosphor screen and sends the data to a computer for analysis.

10.7.3   Conservation of Energy

Fundamental theory is used to analyze the images obtained in a VMI experiment. During a photodissociation event, part of the energy initially contained in the photodissociating photon ($E = h\nu$) is used to break the chemical bond, with dissociation energy, D_0. The excess energy necessarily must go into either internal energy (E_{int}) of the resulting fragments (electronic, vibrational, rotational) or translational kinetic energy (E_T).

$$h\nu - D_0 = E_{int} + E_T \qquad (10.23)$$

VMI experiments directly measure E_T using the time-of-flight from the point of photodissociation to the surface of the detector, and observe the pathways for the disposal of excess energy ($E_{excess} =$

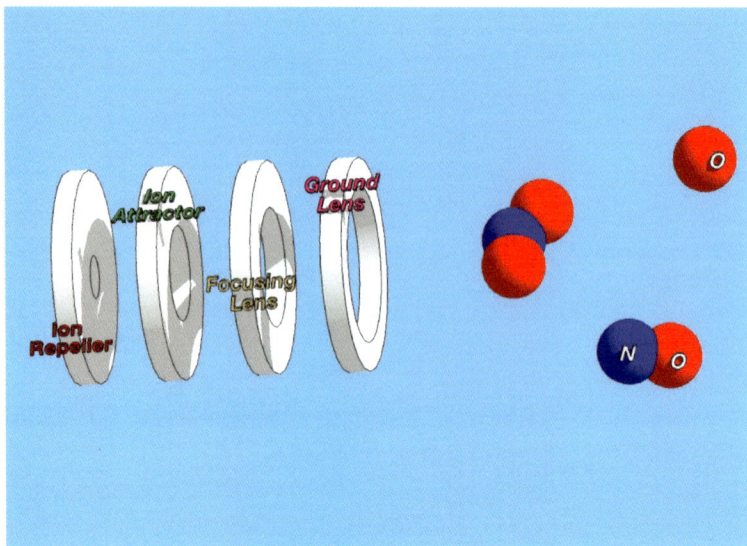

Figure 10.5: Virgo ion optics design for VMI experiments.

$h\nu - D_0$) into internal vibrational energy of photofragments by image analysis. According to the law of conservation of energy, if the photon energy and bond energy is known, then a measurement of the internal and translational kinetic energy of one fragment can be used to determine the amount of energy available to the other partner recoiling photofragment.

10.7.4 Photofragment Angular Distribution

In order to completely describe the dynamics of photodissociation, you need to know three things. We have already discussed how VMI can be used to determine the pathways for energy decay into the product photoframents using the law of conservation of energy. VMI can also be used to determine the photofragment angular distribution and the rate of dissociation. Product angular distributions are important, because the distribution of photofragments as they recoil from the photodissociation region is not isotropic. Instead, the distribution is strongly correlated with the polarization of the dissociating light.

Examine the paper by Chandler *et. al.* [15] that describes the 266 nm photolysis of methyl iodide, CH_3I. The velocity map images in this paper show the 2D projection of speed and angular distribution of the methyl radical products. The recoiling CH_3 fragments are distributed in proportion to $\cos^2 \theta$, where θ is angle between the fragment recoil direction and polarization direction of the dissociating photon. This is exactly the same relationship as in our 3D glasses example in chapter 1!

Relationships like the one between the direction of a recoiling fragment and the direction of polarization of dissociating light are called vector correlations. Vector correlations describe the relationship between the vector properties of the physical phenomena that control a photodissociation process. Specifically, the important vectors are the polarization of the electric field vector, E, of the UV light exciting the parent molecule, the parent molecule's transition dipole moment, μ, the relative velocities of the recoiling photofragments, v, the orbital angular momentum polarization of any atomic photofragment(s) and the rotational angular momentum of any molecular photofragment(s), J. The important question to answer with regard to vector correlations in a velocity map imaging experiment is, what is the velocity vector correlation between the direction of linear polarization of the dissociating laser light and the recoiling photofragments?

In general, for a one-photon photodissociation, the photofragment angular distribution is:

$$I(\theta) = \frac{\sigma}{4\pi}[1 + \beta P_2(\cos \theta)] \qquad (10.24)$$

where $I(\theta)$ is the angular scattering distribution of the photofragments, θ is the angle between μ and E, σ is the cross section for dissociation, $P_2(\cos \theta) = \frac{1}{2}(3 \cos^2 \theta - 1)$ is second Legendre polynomial, and $\beta = 2P_2(\cos \chi)$ is the anisotropy parameter where χ is the angle between μ and recoil direction v. The sign and magnitude of β can be related to the orientation of the transition dipole moment, μ, in the parent molecule, the lifetime of the molecule in the excited state and the symmetry of the excited state. The lifetime of the molecule in the electronically excited state can be determined, because if the lifetime is short, the photofragments may have a strong vector correlation with the polarized light. If the lifetime is long compared to the rotational period of the molecule, the rotation will scramble the alignment.

Chapter 11

Conclusion and Future Outlook

You have learned that quantum mechanics is the basis for important modern technology that is widely used on a daily basis. In making a connection between the pure mathematics of quantum mechanics and technology, you have also seen that in studying quantum mechanics, one also learns the basic principles of vectors, determinants, matrices and angular momentum. These are all mathematical subjects that every student of science should be exposed to, and have applicability to chemistry, physics, biology and engineering.

The quantum nature of matter can be found everywhere, from the far reaches of our universe to your living room. You've read that quantum mechanics has relevance in the common household, and is not just some exotic subject only discussed by scientists working on billion-dollar atom smashers on a quest to discover the secrets of the Big Bang. The field of radioastronomy is built upon the hyperfine transition in the ground state of the hydrogen atom. Much of what we know about our universe comes from radio astronomy. Technology used every day such as GPS devices in your car or on your phone is based on the atomic clock. Atomic clocks would not be possible without the hyperfine structure of atoms such as cesium. The future generation of computers may involve quantum computing. Now you can gain entry into understanding of how these devices might work. The strange and fuzzy nature of quantum mechanics forces us to think differently about how things behave at the atomic and molecular level.

The quantum connection between matter and light is very useful in the development of modern technology. Classical mechanics fails to explain how a hot object emits light, but quantum theory describes this phenomenon with great accuracy. Your HD TV, compact fluorescent lights and lasers wouldn't exist without quantum mechanics. Lasers are quantum devices that take advantage of the quantized nature of matter to produce coherent light. They are used in fundamental science to understand chemical dynamics, in airport security for detection of hazardous materials and at almost every checkout counter for reading bar codes. Lasers have revolutionized medicine through the use of fluorescent probes in imaging tissue and detection of disease.

Quantum mechanics helps us understand atmospheric photochemistry, from the elegant ozone cycle in the stratosphere to pollution in the troposphere. The beauty of the detailed mechanism and dynamics of atmospheric chemistry can be elucidated using laser-based photodissociation and imaging techniques.

You should now continue to be curious about how quantum mechanics comes into play in other disciplines and developing technology, and be confident that you have the basic skills to understand why quantum mechanics is vital and relevant to life in our modern society.

Bibliography

[1] Anon. New radio waves traced to centre of the Milky Way. *The New York Times*, 82:1, May 1933.

[2] P. Atkins and R. Friedman. *Molecular Quantum Mechanics*. Oxford University Press, New York, 4th edition, 2008.

[3] P.F. Bernath. *Spectra of Atoms and Molecules*. Oxford University Press, Inc., New York, NY, 2nd edition, 2005.

[4] K.L. Bittinger, W.L. Virgo, and R.W. Field. Spectral signatures of inter-system crossing mediated by energetically distant doorway levels: Examples from the acetylene S_1 state. *J. Phys. Chem. A Feature Article*, 115(43):11921–11943, September 2011.

[5] N. Bohr. On the constitution of atoms and molecules, Part I. *Philos. Mag.*, 26:1–25, 1913.

[6] N. Bohr. On the constitution of atoms and molecules, Part II. *Philos. Mag.*, 26:476–502, 1913.

[7] N. Bohr. On the constitution of atoms and molecules, Part III. *Philos. Mag.*, 26:857–875, 1913.

[8] M. Born. Statistical interpretation of quantum mechanics. *Science*, 122(3172):675–679, October 1955.

[9] D.M. Brink and G.R. Satchler. *Angular Momentum*. Oxford University Press, New York, 3rd edition, 1999.

[10] J.M. Brown and A. Carrington. *Rotational Spectroscopy of Diatomic Molecules*. Cambridge University Press, New York, 2003.

[11] D. Budker, D.F. Kimball, and D.P. DeMille. *Atomic Physics: An Exploration Through Problems and Solutions*. Oxford University Press, New York, NY, 2004.

[12] R. Campargue, editor. *Atomic and Molecular Beams: The State of the Art 2000*. Springer, New York, 2001.

[13] F. Capra. *The Tao of Physics*. Shambhala Publications, Inc., Boulder, CO 80302, 1975.

[14] D.W. Chandler and P.L. Houston. Two-dimensional imaging of state-selected photodissociation products detected by multiphoton ionization. *J. Chem. Phys.*, 87:1445, 1987.

[15] D.W. Chandler, J.W. Thoman, M.H.M. Janssen, and D.H. Parker. Photofragment imaging: The 266nm photodissociation of CH_3I. *Chem. Phys. Lett.*, 156(2,3):151–158, March 1989.

[16] B.-Y. Chang, R.C. Hoetzlein, J.A. Mueler, J.D. Geiser, and P.L. Houston. Improved 2D product imaging: The real-time ion-counting method. *Rev. Sci. Instrum.*, 69:1665, 1998.

[17] C.W. Chou, D.B. Hume, J.C.J. Koelemeij, D.J. Wineland, and R. Rosenband. Frequency comparison of two high-accuracy Al^+ optical clocks. *Phys. Rev. Lett.*, 104:070802, February 2010.

[18] M.P. Minitti D. Townsend and A.G. Suits. Direct current slice imaging. *Rev. Sci. Instrum.*, 74(4):2530, 2003.

[19] P.C.W. Davies and D.S. Betts. *Quantum Mechanics*. Number 8 in Physics and Its Applications. Chapman & Hall, New York, 2nd edition, 1994.

[20] L. de Broglie. *Researches on the quantum theory*. PhD thesis, Paris University, Paris, France, 1924.

[21] R.F. Delmdahl, B.L.G. Bakker, and D.H. Parker. Completely inverted ClO vibrational distribution from OClO ($^2A_2 24, 0, 0$). *J. Chem. Phys.*, 112:5298, 2000.

[22] A.R. Edmonds. *Angular Momentum in Quantum Mechanics*. Princeton University Press, Princeton, NJ, 3rd edition, 1974.

[23] A. Einstein. On a heuristic point of view about the creation and conversion of light. *Annalen der Physik*, 17:132–148, 1905.

[24] J.E. Elsila, D.P. Glavin, and J.P. Dworkin. Cometary glycine detected in samples returned by Stardust. *Meteoritics and Planetary Science*, 44(9):1323–1330, 2009.

[25] A.T.J.B Eppink and D.H. Parker. Velocity map imaging of ions and electrons using electrostatic lenses: Application in photoelectron and photofragment ion imaging of molecular oxygen. *Rev. Sci. Instrum.*, 68:3477, 1997.

[26] H.I. Ewen and E.M. Purcell. Observation of a line in the galactic radio spectrum. *Nature*, 168:356, June 1951.

[27] Richard P. Feynman, Robert B. Leighton, and M. Sands. *The Feynman Lectures on Physics: Quantum Mechanics*, volume 1. Addison-Wesley Publishing Company, Reading, MA, 3rd edition, July 1966.

[28] B. Friedrich and D. Herschbach. Stern and Gerlach: How a bad cigar helped reroient atomic physics. *Phys. Today*, 56(12):53–59, 2003.

[29] H. Goldstein. *Classical Mechanics*. Addison-Wesley Series in Physics. Addison-Wesley Publishing Company, Reading, MA, 2nd edition, July 1981.

[30] M. Gupta and D. Herschbach. A mechanical means to produce intense beams of slow molecules. *J. Phys. Chem. A*, 103:10670–10673, November 1999.

[31] M. Gupta and D. Herschbach. Slowing and speeding molecular beams by means of a rapidly rotating source. *J. Phys. Chem. A*, 105:1626–1637, October 2001.

[32] J. Hecht. *Understanding Lasers: An Entry Level Gude*. IEEE Press, New York, NY, 2nd edition, 1992.

[33] G. Herzberg. *Molecular Spectra and Molecular Structure II. Infrared and Raman Spectra of Polyatomic Molecules*, volume II. D. Van Nostrand Company, Princeton, NJ, 1945.

[34] G. Herzberg. *Molecular Spectra and Molecular Structure I. Spectra of Diatomic Molecules*, volume I. D. Van Nostrand Company, Princeton, NJ, 1950.

[35] G. Herzberg. *Molecular Spectra and Molecular Structure III. Electronic Spectra and Electronic Structure of Polyatomic Molecules*, volume III. D. Van Nostrand Company, Princeton, NJ, 1966.

[36] Apple Inc. Pages.

[37] Google Inc. SketchUp.

[38] P.M. Johnson. Molecular multiphoton ionization spectroscopy. *Acc. Chem. Res.*, 13:20–26, 1980.

[39] P.M. Johnson and C.E. Otis. Molecular multiphoton spectroscopy with ionization detection. *Ann. Rev. Phys. Chem.*, 32:139–157, 1981.

[40] T.F. Jordan. *Quantum Mechanics in Simple Matrix Form.* Dover Publications, Inc., Mineola, NY, 1986.

[41] Y. Liu K.-C. Lau and L.J. Butler. Photodissociation of 1-bromo-2butene, 4-bromo-1-butene, and cyclopropylmethyl bromide at 234nm studied using velocity map imaging. *J. Chem. Phys.*, 125:144312, 2006.

[42] S. Knappe, V. Shah, P.D.D. Schwindt, L. Hollberg, J. Kitching, L.A. Liew, and J. Moreland. A microfabricated atomic clock. *Appl. Phys. Lett.*, 85:1460–1462, July 2004.

[43] H. Lefebvre-Brion and R.W. Field. *The Spectra and Dynamics of Diatomic Molecules.* Elsevier Academic Press, Boston, MA, 2004.

[44] F.S. Levin. *An Introduction to Quantum Theory.* Cambridge University Press, Cambridge, UK, 2002.

[45] S. Li, J. Matthews, and A. Sinha. Atmospheric hydroxyl radical production from electronically excited NO_2 and H_2O. *Science*, 319:1657–1660, 2008.

[46] M.L. Lipciuc and M.H.M. Janssen. High-resolution slice imaging of quantum state-to-state photodissociation of methyl bromide. *J. Chem. Phys.*, 127:224310, 2007.

[47] L. Lorini, N. Ashby, A. Brusch, S. Diddams, R. Drullinger, E. Eason, T. Fortier, P. Hastings, T. Heavner, D. Hume, W. Itano, S. Jefferts, N. Newbury, T. Parker, T. Rosenband, J. Stalnaker, W. Swann, D. Wineland, and J. Bergquist. Recent atomic clock comparisons at NIST. *Eur. Phys. J. Spec. Top.*, 163:19–35, 2008.

[48] A.D. Ludlow, T. Zelevinsky, G.K. Campbell, S. Blatt, M.M. Boyd, M.H.G. de Miranda, M.J. Martin, J.W. Thomsen, S.M. Foreman, J. Ye, T.M. Fortier, J.E. Stalnaker, S.A. Diddams, Y.L. Coq, Z.W. Barber, N. Poli, N.D. Lemke, K.M. Beck, and C.W. Oates. Sr lattice clock at 1×10^{-16} fractional uncertainty by remote optical evaluation with a Ca clock. *Science*, 319(5871):1805–1808, March 2008.

[49] F.G. Major. *The Quantum Beat: Principles and Applications of Atomic Clocks*. Springer Science+Business Media, LLC, New York, NY, 2nd edition, 2007.

[50] J.C. Mather. Calibrator design for the COBE far infrared absolute spectrophotometer (FIRAS). *Astrophys. J.*, 512:511, 1999.

[51] J. Mehra and H. Rechenberg. *The Historical Development of Quantum Theory: The Formulation of Matrix Mechanics and Its Modifications*, volume 3. Springer-Verlag, New York, NY, 1982.

[52] Albert Messiah. *Quantum Mechanics*. Dover Publications, Inc., Mineola, NY, 1999.

[53] R.L. Miller, A.G. Suits, P.L. Houston, R. Toumi, J.A. Mack, and A.M. Wodtke. The ozone deficit problem: $O_2(X,v>26)$ + $O(^3P)$ from 226-nm ozone photodissociation. *Science*, 265:1831, 1994.

[54] C. G. Morgan, M. Drabbels, and A. M. Wodtke. The correlated product state distribution of ketene photodissociation at 308 nm. *J. Chem. Phys.*, 104:7460, 1996.

[55] M. D. Morse. Supersonic beam sources. *Exper. Meth. Phys. Sci.*, 29B:21, 1996.

[56] NASA. The multiwavelength Milky Way, September 2010.

[57] NRAO. National Radio Astronomy Observatory, June 2011.

[58] W.H. Oskay, S.A. Diddams, E.A. Donley, T.M. Fortier, T.P. Heavner, L. Hollberg, W.M. Itano, S.R. Jefferts, M.J. Delaney, K. Kim, F. Levi, T.E. Parker, and J.C. Bergquist. A single-atom optical clock with high accuracy. *Phys. Rev. Lett.*, 97:020801, July 2006.

[59] L. Pauling and E.B. Wilson. *Introduction to Quantum Mechanics with Applications to Chemistry.* Dover Publications, Inc., Mineola, NY, 1963.

[60] M. Planck. On the law of the energy distribution in the normal spectrum. *Ann. Phys.*, 309(3):553–563, 1901.

[61] R. McWeeny. *Quantum Mechanics: Principles and Formalism.* Dover Publications, Inc., Mineola, NY, 2003.

[62] M.V. Ivanov R. Schinke, S.Yu. Grebenshchikov and P. Fleurat-Lessard. Dynamical studies of the ozone isotope effect: a status report. *Annu. Rev. Phys. Chem.*, 57:625, 2006.

[63] N.F. Ramsey. *Molecular Beams.* Oxford University Press, New York, 1990.

[64] J.S. Rigden. *Hydrogen: The Essential Element.* Harvard University Press, Cambridge, MA, 2003.

[65] M.E. Rose. *Elementary Theory of Angular Momentum.* John Wiley & Sons, Inc., Mineola, NY, 1957.

[66] T. Rosenband, D. B. Hume, P. O. Schmidt, C. W. Chou, A. Brusch, L. Lorini, W. H. Oskay, R. E. Drullinger, T. M. Fortier, J. E. Stalnaker, S. A. Diddams, W. C. Swann, N. R. Newbury, W. M. Itano, D. J. Wineland, and J. C. Bergquist. Frequency ratio of Al^+ and Hg^+ single-ion optical clocks; metrology at the 17th decimal place. *Science*, 319(5871):1808–1812, March 2008.

[67] R. Schinke. *Photodissociation Dynamics*. Cambridge University Press, Cambridge, UK, 1993.

[68] J.T. Schwartz. *Introduction to Matrices and Vectors*. Dover Publications, Inc., Mineola, NY, 2001.

[69] P.D.D. Schwindt, S. Knappe, V. Shah, L. Hollberg, J. Kitching, L.A. Liew, and J. Moreland. Chip-scale atomic magnetometer. *Appl. Phys. Lett.*, 85:6409–6412, October 2004.

[70] G. Scoles, editor. *Atomic and Molecular Beam Methods*, volume 1 and 2. Oxford University Press, New York, 1988.

[71] O. Shimomura. The discovery of aequorin and green fluorescent protein. *J. Microsc.*, 217:3–15, 2005.

[72] T.C. Steimle and W. Virgo. The permanent electric dipole moments for the $A^2\Pi$ and $B^2\Sigma^+$ states and the hyperfine interactions in the $A^2\Pi$ state of lanthanum monoxide, LaO. *J. Chem. Phys.*, 116:6012, 2002.

[73] T.C. Steimle and W. Virgo. The permanent electric dipole moments and magnetic hyperfine interaction in the $A^2\Pi$ state of yttrium monosulfide. *J. Mol. Spec.*, 221:57, 2003.

[74] T.C. Steimle and W. Virgo. The permanent electric dipole moments and magnetic hyperfine interactions of ruthenium mononitride, RuN. *J. Chem. Phys.*, 119(24):12965, 2003.

[75] T.C. Steimle and W. Virgo. The permanent electric dipole moments of the $X^3\Delta$, $E^3\Pi$, $A^3\Phi$ and $B^3\Pi$ states of titanium monoxide, TiO. *Chem. Phys. Lett.*, 381:30, 2003.

[76] T.C. Steimle and W.L. Virgo. The permanent electric dipole moments of WN and ReN and nuclear quadrupole interaction in ReN. *J. Chem. Phys.*, 121(24):12411, 2004.

[77] T.C. Steimle, W.L. Virgo, and J.M. Brown. Permanent electric dipole moments and hyperfine interaction in ruthenium monocarbide, RuC. *J. Chem. Phys.*, 118(6):2620, 2003.

[78] T.C. Steimle, W.L. Virgo, and D.A. Hostutler. The permanent electric dipole moments of iron monocarbide, FeC. *J. Chem. Phys.*, 117:1511, 2002.

[79] T.C. Steimle, W.L. Virgo, and T. Ma. The permanent electric dipole moment and hyperfine interaction in ruthenium monofluoride (RuF). *J. Chem. Phys.*, 124:024309, 2006.

[80] W. Stein and D. Joyner. SAGE: System for algebra and geometry experimentation. *Comm. Computer Algebra*, 39:61–64, 2005.

[81] J.I. Steinfeld. *Molecules and Radiation.* Dover Publications, Inc., Mineola, NY, 2nd edition, 1985.

[82] A.G. Suits and R.E. Continetti, editors. *Imaging in Chemical Dynamics.* ACS Symposium Series 770. American Chemical Society, Washington, DC, 2001.

[83] H.H. Telle, A.G. Urena, and R.J. Donovan. *Laser Chemistry: Spectroscopy, Dynamics and Applications.* John Wiley & Sons Ltd., West Sussex, England, 2007.

[84] W.J. Thompson. *Angular Momentum: An Illustrated Guide to Rotational Symmetries for Physical Systems.* John Wiley & Sons, Inc., New York, 1994.

[85] S.Y.T. van de Meerakker, N. Vanhaecke, and G. Meijer. Stark deceleration and trapping of OH radicals. *Annu. Rev. Phys. Chem.*, 57:159–190, December 2006.

[86] W.L. Virgo. *Construction of a Continuous Wave Cavity Ringdown Laser Spectrometer for the Gas-Phase Ultratrace Absorption Measurement of Ammonia.* W.L. Virgo Princeton University Thesis, Princeton, NJ, May 2000.

[87] W.L. Virgo. *The Response of Diatomic Chemical Intermediates to Electric and Magnetic Fields.* Thesis, Arizona State University, Tempe, AZ, December 2005.

[88] W.L. Virgo. *The Equations of Chemistry.* Rhodium Inc., Cambridge, MA, 1st edition, 2013.

[89] W.L. Virgo. Simultaneous Stark and Zeeman effects in atoms with hyperfine structure. *Am. J. Phys.*, 81(12):936–942, 2013.

[90] W.L. Virgo, T.C. Steimle, L.E. Aucoin, and J.M. Brown. The permanent electric dipole moments of ruthenium monocarbide in the $^3\Pi$ and $^3\Delta$ states. *Chem. Phys. Lett.*, 391:75, 2004.

[91] W.L. Virgo, T.C. Steimle, and J.M. Brown. Optical Zeeman spectroscopy of the (0,0) bands of the $B^3\Pi$ and $X^3\Delta$ transitions of titanium monoxide, TiO. *Astrophys. J.*, 628:567, 2005.

[92] H. Wang, W.L. Virgo, J. Chen, and T.C. Steimle. Permanent electric dipole moment of molybdenum carbide. *J. Chem. Phys.*, 127:124302, 2007.

[93] C.E. Wayne and R.P. Wayne. *Photochemistry.* Oxford Chemistry Primers. Oxford University Press, New York, NY, 1996.

[94] K.-C. Lau Y. Liu and L.J. Butler. Photodissociation of cyclobutyl bromide at 234nm studied using velocity map imaging. *J. Phys. Chem. A.*, 110:5379, 2006.

[95] N. Yamakita, S. Iwamoto, and S. Tsuchiya. Predissociation of excited acetylene in the A^1A_u state around the adiabatic dissociation threshold as studied by LIF and H-atom action spectroscopy. *J. Phys. Chem. A*, 107(15):2597–2605, 2003.

[96] X. Yang and K. Liu, editors. *Modern Trends in Chemical Reaction Dynamics: Experiment and Theory.* Parts I and II. World Scientific, New Jersey, 2004.

[97] R.N. Zare. *Angular Momentum: Understanding Spatial Aspects in Chemistry and Physics.* John Wiley & Sons, Inc., 1988.

[98] G. Zukav. *The Dancing Wu Li Masters: An Overview of the New Physics.* Bantam Books, New York, 1984.

Index

Mercury clock, 78
Metastable, 85
Molecular beam, 90

Normalization, 21

Observable, 21
Operator, 21
Orthogonal, 24
Orthonormality, 24
Overlap integral, 22
Oxygen, 104
Ozone, 104

Pauli, 15
Perturbations, 33
Phosphorescence, 109
Photodissociation, 110
Photoionization, 113
Planck's quantum hypothesis, 2
Polarization, 10
Population inversion, 88
Postulates of quantum mechanics, 20
Projection, 27, 50

Quantum logic clock, 78, 82
Quartz oscillator, 76
Qubit, 81

Radar, 79
Radio astronomy, 71
Rotation matrices, 35
Rovibronic, 89, 115
Rubidium clock, 77

Scalar product, 28
Schrödinger equation, 36, 42
Secular determinant, 32
Secular equation, 34
Similarity transformation, 33

Singlet state, 53, 84
Spectroscopy, 90
Spin-orbit, 66, 85
Stark, 65
Stern-Gerlach experiment, 90
Stimulated emission, 87, 88
STM, 16
Strontium clock, 77

Triplet state, 53, 84

Uncoupled representation, 51, 58, 64, 69
Unit matrix, 30

Vector representation, 26

Wave superposition, 22
Wave-particle duality, 12
Wavefunction, 20
Wigner 3j symbol, 52

Zeeman, 59

Printed in Great Britain
by Amazon

80434957R00080